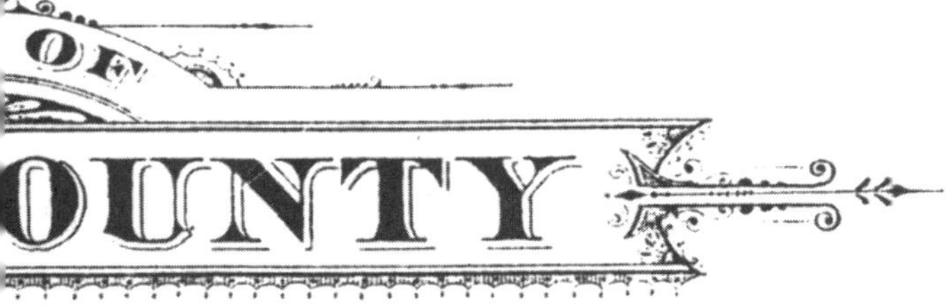

OIS.

Russell Sta.

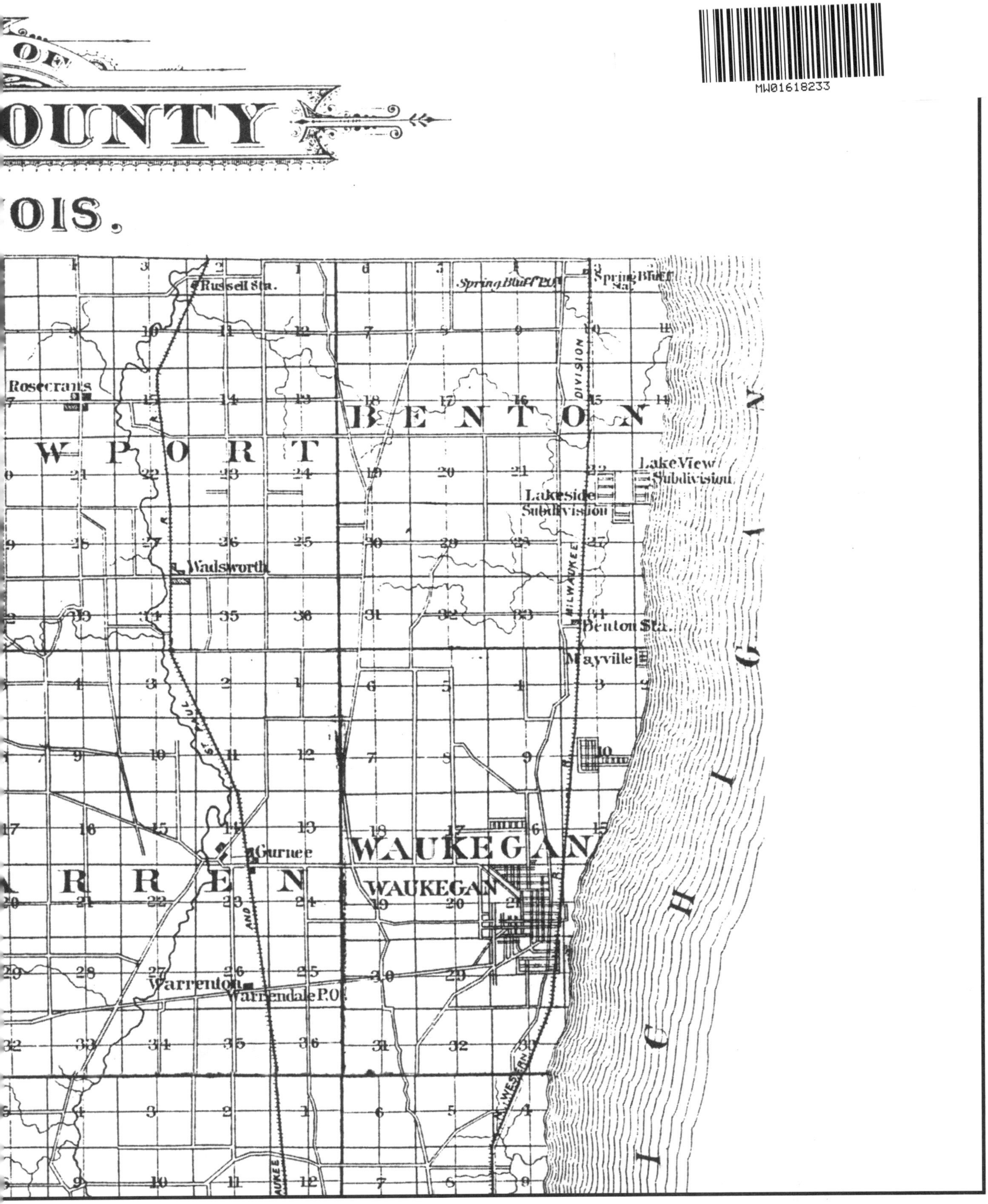

Spring Bluff P.O.

Spring Bluff Sta.

Rosecrans

B E N T O N

W P O R T

Lake View Subdivision

Lakeside Subdivision

DIVISION

Wadsworth

MILWAUKEE

Benton Sta.

Mayville

ST. PAUL

Gurnee

WAUKEGAN

A R R E N

WAUKEGAN

AND

Warrenton

Warrendale P.O.

WESTERN

I C H I G A N

Land Conservancy of Lake County

The Land Conservancy of Lake County (LCLC) is proud to sponsor *The Barns of Lake County.*

LCLC was founded in 1989 by a group of environmental professionals and concerned citizens as a land trust that protects and manages natural and open lands in Lake County, Illinois. LCLC is the oldest countywide land trust, with expertise in managing both large and small properties through volunteer stewardship. As a supporting member of the national Land Trust Alliance, LCLC follows its guidelines for the responsible and ethical operation of land trusts.

More information about the LCLC can be found on the web at www.landconservancyoflakecounty.org.

The Barns of Lake County

Text and photographs by Nancy Schumm-Burgess

THE
DONNING COMPANY
PUBLISHERS

DEDICATION

This book is dedicated to my children, Amanda and Patrick. May their memories of our forays into the barns of Lake County be good ones.

For information write:
The Donning Company Publishers
184 Business Park Drive, Suite 206
Virginia Beach, VA 23462

Steve Mull, General Manager
Ed Williams, Project Director
Barbara Buchanan, Office Manager
Julia Kilmer, Editor
Pamela Koch, Editor
Lori Wiley, Senior Graphic Designer
Scott Rule, Director of Marketing
Travis Gallup, Marketing Coordinator

All photographs in *The Barns of Lake County* were taken between April 1997 and November 2000 by Nancy Schumm-Burgess.

Library of Congress Cataloging-in-Publication Data:
Schumm-Burgess, Nancy, 1963–
The barns of Lake County / text and photographs by Nancy Schumm-Burgess.
p. cm.
Includes bibliographical references and index.
ISBN 1-57864-225-6 (hc : alk. paper)
1. Barns--Illinois--Lake County--History--19th century.
2. Barns--Illinois--Lake County--History--20th century. I. Title.

NA8230.S385 2003
728'.922'0977321--dc22

2003055480

Printed in the United States of America at Walsworth Publishing Company

Photos by Kate Roth © 1998

TABLE OF CONTENTS

OLD BARN

Good morning old barn across the way
I love to see you day after day.
You're my symbol of strength and tranquility,
Part of each sunset that brings peace to me.
Thru the years you must have had stories to tell
Of when you loved and sheltered all animals well,
Stored hay in the lofts and grain in the bin.
There was always plenty of food and a place to rest when
The swallows and bats, after their all nite roam,
Would fly back to your rafters and call it their home.
Your foundations may crumble and you will fall in a heap,
But my thoughts and memories I'll always keep
For I love the many splendored things,
Like majestic silence a shabby old barn brings.

—Agnes De Vries Dudink
Lake County, IL
94 years
1998

INTRODUCTION

IF WE ARE LUCKY, WE ALL HAVE MOMENTS OF CLARITY in our lives—moments when our paths stretch neatly before us as if an unknown force has directed us where to go. So it was, when I walked into the first barn loft while taking photographs of the barns in Lake County.

Prior to that time, I had been happily viewing the barns from afar, taking photos from the roadways, never really connecting to the barn owners. It was inevitable that I would enter the barns and meet the people who cherish them. It was these meetings, and my first entrance into the great barn loft, that transformed my purpose from taking photos to the desire to preserve and protect these history-laden relics.

I happened upon this particular barn while I was cruising the back roads in Mundelein. I had been talking to people, anyone who would listen really, about my barn project and they were all lamenting the fate of these barns. I stopped the car when I saw this barn nestled behind an old farmhouse. It was autumn and the barn had become more visible because the trees were beginning to lose their leaves, exposing the world beyond. Knocking on the farmhouse door, I interrupted a young woman feeding her baby, with a toddler at her heels. I gave her my spiel, how I was writing a book and could I take some photos. She agreed that it would be all right.

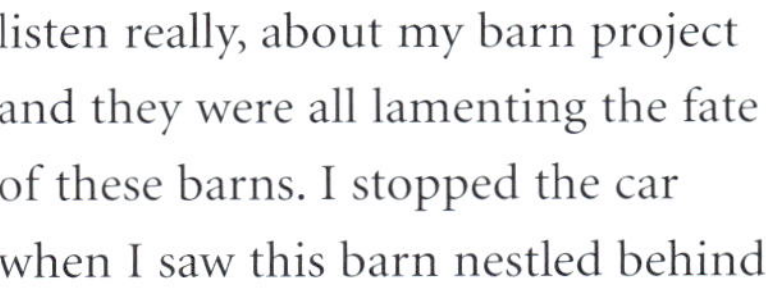

It was not clear how to access the barn, so I walked down along a path to the lower byre. I climbed over an old fence and admired the crumbling stone foundation indicative of German builders. Noises of animals scurrying for cover could be heard. There was an eerie stillness to the lower level, an echo of the livestock that had once lived there and the families that had raised them. I walked into an area that had been added in later years of the barn's life and saw the stanchions, the milk cooling room, and stalls where workhorses had resided. I could envision the previous life of the old barn, and I drifted into a reverie of imagined activity.

I shook myself from this dream state. I wanted to get a photo of the barn with the trees, but without any vehicles in front, so I ventured into the field to the rear of the barn. This side was somewhat overrun by weeds, and, as I pride myself on my ability to wear the wrong shoes for every occasion, I decided that weeds were not going to discourage me. I forged through the short tangle, which became taller, more prickly, and denser as the ground sloped away from the barn.

Finally, after about fifty feet of struggling through the tangled overgrowth, the weeds cleared and the recently mown fields spread before me. There was a well-worn dirt road stretching through the fields, snaking to the other farms spread afar. The effect was breathtaking. With the barn behind me, the fields and dirt road before me, and the hawks circling above calling to each other, I felt as if I had stepped into another time. Did a place like this really still exist in the midst of our advancing urban sprawl? It was the first time that I was struck by how well farmers chose the best spots to set their barns.

I took my photos and followed the road back to the barn (a much more efficient, less tangled path). Plucking the burrs and seeds from my sweater, I walked up to the threshing floor on the middle level of the barn and found myself overwhelmed again for the second time that day.

This view of the barn was entirely new for me. As I gazed beyond the castaway junk of our modern era at the rough-hewn logs, the hand-hewn beams, and the massive expanse of the loft, I fell in love. Sun rays streamed through the loose siding casting surreal streams of light on the old beams. Upon the left haymow was a date burned into the wall in the Olde English writing style: 1877. Just then I decided that I would do something to save these places.

There is, I have since discovered, in even the most crumbling of structures, a certain spiritual essence to barn lofts. Some are clearly cathedral-like, not only in the ethereal way that light filters through the cracks of old wood, but also in the way that all of one's senses are stimulated. The aged wood carries a certain scent of old hay and forests. Ears are treated to the creaks and groans of the loose boards and the sounds of animals rustling. The eyes adjust to the lowered lighting and spaces in the brain seldom reached. One is touched by how rare this sense of peace is in our bustle of modern living.

To understand that a handful of men worked together to build each of these barns, and the hopes that each barn represented, makes them priceless. The barn is one of the most underrated structures known to modern man. But, the barn's history shows that it was the most precious structure on a farm. More valuable than even the home, the barn protected the animals, food stores, equipment, and most importantly, the grain and seed that made the farm a viable business. Barns were indeed "prairie castles."

At the turn of the last century, there stood as many as 2,300 barns in Lake County. Today, about 240 remain. The people who own the barns now love them, even if they cannot afford to maintain them. Neighbors who live around the barn feel that it belongs to them, just as much as to the family that owns the property.

The goal of this book is to help the barns and farmland that remain. Even barns that have been rehabilitated possess a spirit of those who built them. Saving them means capturing that spirit to share with future generations. If we do not maintain the historically significant barns, the only barns the next generation will know will be contrived versions of the truth.

It is with love that this book was created, and with hope that it will help to save at least some of the "prairie castles" before they disappear forever.

—Nancy Schumm-Burgess, 2004

1

THE PIONEERS

IN 1673, LOUIS JOLIET DESCRIBED IN HIS JOURNALS the richness of the Illinois prairie soil: "At first, when we were told of these treeless lands, I imagined that it was a country ravaged by fire, where the soil was so poor that it could produce nothing. But . . . no better soil can be found. . . . A settler would not there spend ten years in cutting down and burning the trees; on the very day of his arrival, he could put his plough into the ground."

With these words from Joliet, it was only a matter of time before those who had come to the Americas for a new life would head as far west as Illinois. In fact, the new Americans would fight many battles to secure ownership of this area. In 1833 the last Treaty of Chicago was signed, calling for permanent removal of the Native Americans, and the land was officially opened for settlement.

While the federal government determined that no one could technically lay claim to the land in Lake County before autumn of 1836, homesteaders had put down stakes as early as 1834. The pioneers saw the land potential stretching before them and eagerly claimed as much as they could to secure their families' futures. The Pre-Emption Law of 1819 stated that settlers could hold legal claim to the land before it was surveyed. The claimants lived on the land and farmed as they saw fit. They marked the borders of their property by laying a furrow, or felling trees to mark the boundaries (according to Edward Lawson who wrote the early history of Warren Township). It was only after the official surveys, some of which were completed as early as 1837 and others as late as 1844 in Illinois, that land could be paid for and a deed filed. When the federal land offices opened in Chicago for purchases, the land was cheap, only $1.25/acre.

The original deeds were issued directly by President John Tyler, who was in office from 1841 until 1845, and then from President James Polk who encouraged the expansion to the west through federal land grants. There are a few families today who still hold these original land grants.

Cuba Township. This English-style barn was built by the White family. The Whites also constructed the first schoolhouse in the area, and they established the White Cemetery. The barn was built using one main massive beam called a swing beam that was the main support for the post-and-beam construction of the entire barn. Swing beams were indicative of very early English settlement barns. Several additions have been made to this barn over the last 150 years. The lower level and soil-banked foundation were built during one addition, while another addition to the barn included stalls for horses.

The northern portion of Illinois was settled primarily by German and English settlers who had come from New York, New England, and some from Pennsylvania. They arrived by overland trails, or through the Erie Canal and the Great Lakes. This influx of northerners or "Yankees" ensured that the state of Illinois would be a "free" state. They followed the traditions of the New England settlements by enacting the township form of government, creating the county boundaries, and legally establishing Lake County on March 1, 1839.

Pioneers were resourceful and dedicated to improving life in the county. In 1840, there were 2,905 residents in Lake County. As word spread of the fertile soil, ample supply of hardwood, and availability of fresh water, more pioneers arrived. The dangers of cholera in the 1840s sent people to the outlying areas of Chicago. These early residents expressed their concerns about education and commerce, and they brought about the establishment of country schools and community businesses. Churches were formed and wagon shops, blacksmiths, and taverns were built for the growing communities.

Every man who came here was a farmer first, out of necessity. Those early families brought with them provisions, seed, and some livestock to be ready for the troubles that might lie ahead. They housed the family first with log cabins or sod homes, and then prepared the wood necessary to build barns. The barns would protect their food stores, grain, and livestock from native predators like wolves, coyotes, lynx, wildcats, and weasels.

This sketch illustrates the framework of an English wheat, self-standing, three-bay design barn.

Top: *Framework sketch of a German bank barn with soil banking.*

There was food aplenty in this new frontier. Crops and livestock were grown and raised to support the family throughout the coming year. Early crops produced were potatoes, corn, wheat, and oats. Vegetables for storage included pumpkins, carrots, turnips, parsnips, and beets. There were also wild fruits like strawberries, blackberries, raspberries, cherries, crabapples, and gooseberries for the picking. Wild grapes helped to flavor the wild game which was plentiful for the early settlers—pigeons, deer, raccoon, rabbit, squirrel, duck, goose, quail, partridge, and prairie chicken, and fish from all the fresh-water tributaries. Eventually domesticated animals were imported. Sheep, hogs, cattle, chickens, ducks, and turkey were bred for meat, lard, wool, and tallow to make candles that were the source of early light.

By the 1840s, improvements came to farmers in the form of agricultural schools and industrial inventions, most notably the invention of the steel plow and the reaper. In 1850, only a few short years since the first settlers struck their claims in 1834, plans were underway to bring the railroads west. This expansion would mean that the supply of materials and products to and from major markets would increase dramatically.

FIRST SETTLEMENT BARNS

The first generation of Lake County settlers needed barns to protect their families' livelihood. The farmhouse was for the family, but the barn was for the business, the "headquarters" of the business, as it were. Barns would hold the grain and seed, farm equipment, and the animals that would feed the family and work the fields. The barns were built by combining new American innovations with German and English barn building traditions to create our early settlement barns.

Cuba Township. This fragment of a barn is the oldest section of barn found to date in the county, having been built around 1834. The legend of this barn says that a German barn builder was traveling through the area and taught a group of Native Americans how to build barns in the traditional German method. The hand-hewn logs and early-style mortar and fieldstone foundation, plus the soil banking on the north side, do indicate that Germans built the barn. But, according to local Native American historians, the likelihood of Native Americans in 1834 cooperating and learning from travelers passing through is very slim. However, Daniel Wright, who is recognized as the first settler of the county in 1834, did document friendly encounters with Native Americans who aided him in building his first dwelling.

American tradition was to set the barn away from the home and closer to the fields where it was needed. Early European tradition was to connect the barn directly to the house or under the house. This kept the house warmer, it was believed, but came with problems like sanitation and odor. Americans corrected these problems by improving ways to warm the house and by moving the barn away from the house and often downwind.

Two barn designs were favored during this early settlement period: the English style and the German style. The English style of barn was designed for wheat production, with animals and other crops separated. Historians claim that the wheat farmers would say that "no self-respecting farmer would combine his dirty animals with his wheat." The German style barn was all inclusive combining animals in the lower level and crops and equipment in the upper level.

Both designs called for the foundation for the barn to be set on the best part of the property, often on the highest point for the greatest drainage. Foundations were created using available materials. Fields were cleared of large stones and the Germans, who were master

Avon Township. The earliest record of the Bonner family's arrival in Lake County was that of John Bonner in 1842. The family owned 211 acres of land in Lake Villa and continued farming it until 1995, when the farm was donated to the Lake County Forest Preserve. The farm's uniqueness lies in the variety of buildings reflecting several generations of family ownership. Each generation built additions to the old barns, thereby maintaining the standards required by the current markets. The earliest section on the barn is in the foreground and dates to the early 1840s.

masons, created intricate foundations with these stones, fitting rocks together like a puzzle. The largest flat stones formed the base for the English-style barn and served as the cornerstone for the foundation.

The German barns were called bank barns because the soil was banked or built into one side, usually the north side. This custom came from early German barns that were built under the home and along the hillsides. Originally called Swisser barns or Schweizer barns after Swiss origin, the barn was later called Pennsylvania German barn and then unilaterally the "German bank barn." In this barn design, the lower level was accessible from the south side so that the animals would be protected from the north winds. Early designs showed the lower level open with an extension over the open side, called the fore-bay or overhang, but in Lake County, extreme weather conditions called for the lower level to be enclosed completely.

The foundation in these bank barns was all masonry and fieldstone. The masonry consisted of mud or sand combined with water to make an early concrete base. Some of the builders were so expert at their craft that no mortise was needed between the fieldstones. When examining the interior of these older barns, the foundations are sometimes two feet thick with the old masonry barely cracking on the

Moraine Township. A smaller barn on an old family farm, this structure was built by the Garling family. The Garling family came to the area before 1850 and helped to establish the Evangelical Lutheran Congregation in Highland Park.

Antioch Township. This barn belonged to Hiram Buttrick who built the first sawmill in the township in 1839. He arrived in the country in 1838 and moved to Waukegan in 1846.

Ela Township. This barn was built by the Morse family around 1847. The Morse family held the original President Tyler land grant to this tract of land. In 1985 the last family member sold the land. Extended bays were added to the original structure of the barn with new generations of family owners.

interior. The central floor of these barns, easily accessible due to the banking, was where hay was stored, as well as wheat waiting to be threshed. The hay storage, or haymow, was on either side of the threshing floor, and there were several trapdoors available to toss the hay down to the animals below.

The original design for the English barn was the three-bay design. This floor plan was considered ideal for the wheat farmer. It called for a central area or threshing floor in the middle of the barn. This threshing floor had a door on each side of the barn, which some call "the door that leads to nowhere." These doors were designed so that after the threshing, during winnowing, the doors could be opened, and the wind would blow through and blow away the chaff, leaving behind the wheat. The wheat was kept in place by a board strategically placed at the doorway entrance called the threshold. On each side of the threshing floor were two equally sized bays—one side for wheat waiting to be threshed and the other for animal feed and bedding.

Ela Township. Another Morse family farm built around the early settlement period. Large groups of families often traveled and settled together. Through the doorway, one can see the post-and-beam construction of the early settlement barn. The mortise and tenon joints, hewn by the farmer, were joined together with wooden pegs. The beams, called hand-hewn, were cleared of bark and squared by hand from local oak trees using a broadaxe. The barn is now used for storage by a local disposal company.

For both English and German settlers, the barn framework was built using native hardwoods. Beams and joints were formed entirely by hand. The construction was called post and beam and required no metal nails, only wooden pegs that joined mortise and tenon joints. Massive timbers, hewn with a broadaxe, formed the main beams, and rough hewn logs formed the supports. The mortise and tenons were prepared in advance of construction, carefully planned with the auger holes drilled. Wood was planked for siding, floorboards, and ceilings. Even the shakes for the roof shingles were hewn by hand.

These early American barns were built by families and neighbors working together. Community exchange of information and ideas helped to design many early barns. The job of the barn owner was to cure the wood, prepare the mortise and tenon joints, and plan the general structure. When the wood was ready, neighbors would come together for a barn raising. They would build the bents, raise the barn, and enjoy a feast prepared by the women. Later, the farmer and his family would finish the siding and roofing. It was a good custom to be friendly with your neighbors for obvious reasons.

Ela Township. Originally owned by the Ernsting family, this barn is an excellent example of an early settlement barn. Built in the tradition of the old German barns, it was soil-banked on the north face, with a southern exposure for the animals in the lower level. Very few windows were a common sight on early barns, believed to discourage bugs, dust, and exposure to the outside elements. The shortage of glass was also a contributing factor in limiting the use of windows.

Roof lines in this area began with the salt box design, longer on the north side and shorter on the south side so that the prairie winds would blow across the roof line, allowing the south side to absorb more heat. Eventually, this design was replaced by a simple gable peak, triangular in shape. The standard measurement for the single level barn was thirty feet from ground to gable peak.

Few windows are found in the early structures, in part because of the scarcity of glass, but also because it was felt that the lack of windows reduced the dust and bugs. Fewer windows also helped protect the barns from the elements, keeping the barn warmer in the winter and cooler in the summer. Barns that have tilted windows near the roof peak are examples of farmers' efficiency. These windows were recycled from the homesteads back east. After the home was built, if there were extra windows, the farmer, not wanting to waste resources, would simply use the windows in the barn, placing them as close to the roof peak as possible to allow light into the upper lofts.

By 1850, the population in Lake County was increasing steadily, and the settlers' children were growing up and needing land. There was a surge of population growth from newcomers who had heard about the rich soil and resources for growing in Illinois. The barns that had served this generation so well were about to become obsolete, as larger, more efficient barns were needed. Most of these early barns were removed completely, added onto considerably, or used as storage sheds.

Ela Township. This small barn was constructed by the Miller family. The supports were created with very rough hewn logs joined with wooden pegs. The roof line is called a salt-box line, considered by early settlers to be a better way to temper the wind that blew fiercely across the prairie.

Ela Township. The Remlinger family, one of the early settlement families on the north side of Long Grove, built this barn. Later the farm was owned by the Butt family who joined together some area buildings to form their farmhouse. One of the buildings used was the first church in Long Grove. The church was converted into the parlor of the home, and children of the family recall that there was always a special reverence reserved for that room in the house.

Ela Township. This barn in Ela Township was built prior to the Civil War by the Sigwalt family who arrived in the area around 1840. Michael Sigwalt served as the first postmaster in Long Grove. The barn possesses the simple gable roof and H-frame construction of early barns.

Fremont Township. This old barn is on property that once shared a dance hall and general store for the Village of Fort Hill. Also on the property is an old schoolhouse moved from the township line dating back to the original settlement. In its later years the schoolhouse was used for housing animals.

Ela Township. A middle 19th century barn owned by the Barbaras family on the early maps. The giant roller doors were introduced around the 1840s in the United States, patterned after the large doors on freight railroad cars. The door-within-a-door feature was created because the large doors were unyielding if they were not kept greased. The effort to open the big doors was not needed for everyday uses.

Fremont Township. Torn down to make way for modern development, this early barn was once owned by the Woolaridge family. The extended bay was added in later years for more storage.

Newport Township. One of the oldest structures in the county, this barn was built by the Alcock family who arrived in Newport Township in the 1840s. They acquired the land under a Tyler land grant through the Pre-Emption Law, which stipulated that a settler could stake his land prior to the county survey and pay for those acres after the survey was finished. Inside the old barn are very rough hewn logs forming the horizontal supports and supporting the floor of the loft. Members of the original family still owned the land as late as 2000.

Vernon Township. This barn is owned by the Long Grove Historical Society for use as a hands-on museum. The barn was built by the Irwin Ruth family sometime around 1847. The barn is typical of early settlement barns in that it has very few windows, a basic gabled roof, natural wood siding, and post-and-beam construction with mortise and tenon joints. The barn swallows' nests were left behind and considered to be good luck by the early settlers. Barn swallows consumed their weight in bugs and were not as messy or destructive as other birds.

Vernon Township. The Redlinger family built this barn early in the history of the county. For many years it served as a wagon shop and tool shop to travelers along Milwaukee Avenue, then called the Chicago Trail. Today the barn functions as an antiques store.

Vernon Township. This barn was built in the early settlement period. The original section was built in the simple English design of a wheat barn with post-and-beam construction and few windows. The barn had several owners during its lifetime who added the extensions, leaving the original structure intact.

Vernon Township. This barn was the last remaining barn on the original Gridley family property. The Gridleys came to Long Grove in 1835 from New York and set up their homestead. Later family members built their farms on adjoining tracts of land. The barn dates from about 1840. The roof line is a common "prairie style" that slopes over the extended crib on the eastern side to protect animals from prairie winds. This style was later abandoned in favor of the simple gable or gambrel roof line. From the eclectic cuts in the siding, it appears that the barn went through many, many changes in its 160 years of existence. Abbott Laboratories in Long Grove holds their offices in another Gridley family farm nearby built around 1900.

Warren Township. The Strang family were some of the first Scottish settlers in the Millburn area. This barn was part of an original family homestead. The Strangs came from Scotland around 1840 and were known for establishing the first post office in Millburn in 1848.

2

THE SECOND GENERATION 1850–1900

THE SECOND HALF OF THE NINETEENTH CENTURY WAS a time of great change all around the country. These changes affected everyone from the president of the United States to the farmers of the Midwestern states. The development of agricultural schools and farming publications, the reign of Queen Victoria, the industrial revolution, the advance of the railroad, the Civil War, and the great Chicago fire all affected the lives of farmers in Lake County. These changes dictated the way that the farmers in Lake County did business, designed their farms, and raised their families.

Farming was the way of life in Lake County in the second half of the 1800s. The original settlers' children took over portions of land or joined in business with their parents. These children were educated in the new frontier and looked to the world for information to help with the farm. In the 1840s the first breed of agricultural schools had cropped up around the nation. These schools studied ways to improve farming methods. Information and advancements from the schools were published in articles and journals designed to help the farmer and deliver modern solutions directly to the farmer's door. Publications such as the *Prairie Farmer Magazine* began printing as early as 1841 and are still being issued today. There were fifty-nine trade publications for farmers by 1850, which touted the latest inventions and the best solutions for farming problems. These publications brought information from around the globe to the country.

Life for farmers was still challenging because Lake County was still vastly rural. Livestock laws needed to be enacted to protect the farmers' herds. These laws encouraged farmers to corral their livestock, rather than let them wander freely. One solution to the wandering was to plant fences of Osage orange trees. The trees grow quickly and have sharp thorns along their branches that look like barbed wire, and some say they were the prototype for the barbed wire that eventually lined property borders. The penalty for disobeying the new laws was a fine of $5/head for sheep or hogs found loose.

Antioch Township. A well-loved family farm.
The cornerstone says the barn was built in 1873.
The large size, numerous windows, and multi-use
function of the barn indicate the second generation age.

Loose domestic animals were not the only problem for farmers. In 1854 a bounty was offered on the head of wolves killed within the southern part of the county. Still largely forested, the sections we know today as Deerfield and Vernon Hills and along the Des Plaines River were infested with wolves. For every adult head $8 would be rewarded, and for every young head the bounty was $1. By 1868 the problem was nearly solved, but a number remained in Benton Township to the north. A great wolf drive was organized in March of that year. Nelson Landon, who owned several large farms around today's Zion, had 350 men closing in on a pack of wolves. Only three were killed; it is estimated that about twenty-five escaped, but the neighbors were on high alert. A wolf pack could do serious damage to a herd of sheep in a short period of time. These were just some of the obstacles the farmers faced regularly.

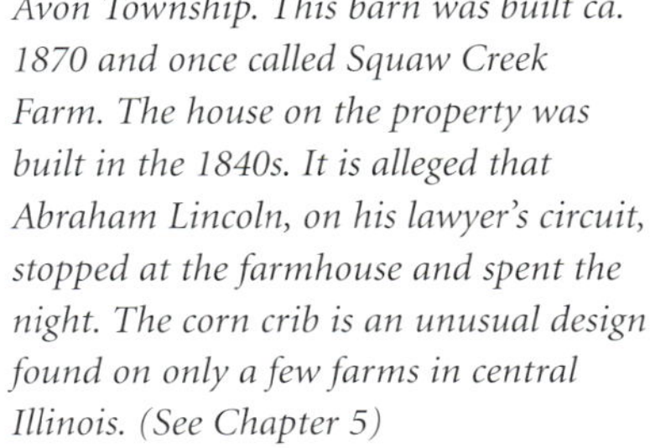

Avon Township. This barn was built ca. 1870 and once called Squaw Creek Farm. The house on the property was built in the 1840s. It is alleged that Abraham Lincoln, on his lawyer's circuit, stopped at the farmhouse and spent the night. The corn crib is an unusual design found on only a few farms in central Illinois. (See Chapter 5)

A large influence in the values of the day was the reign of Queen Victoria (1837–1901). Victoria lived her life by example. She believed strongly in honesty, patriotism, and family values, and these ideals were reflected in the conservatism and patriotism of the day. Farmers were encouraged, through publications and word of mouth, to create

Below: *Cuba Township. Modest in size and possessing few windows, this barn was built around 1860–1870 by the Miller family in the tradition of the early settlers' wheat barn.*

ideal farms. Families believed that the exterior of the house and farm reflected the interior happiness of the family. Farmers who had untidy and unproductive farms were considered to be "lazy farmers."

To help keep the farm in order, inventions that made farming easier were now being produced in the northern states. Cyrus McCormick, who had invented the first successful reaping machine in 1831 in Virginia, opened a factory in Chicago in 1847, bringing his inventions closer to the Lake County farmer. The growth of steel mills provided ironwork for the farms. Not just steel harvesting machines and steel plows, but now wheels, pots, and tools were carried by traveling salesmen who delivered these new inventions directly to the farmers. Inevitably, it was the railroad that brought the large products of the industries right to the farmer's door.

In February of 1851, plans were made to establish the first railroad through Lake County. The initial goal was to stretch the line from Chicago, through Waukegan, to the Wisconsin border. By March of 1851 the plans included Milwaukee and north to Green Bay. Finally, in 1855, the first run was made to Waukegan, and late in the year the run to Milwaukee was completed. All along the route, stops were established, and around these stops small towns of commerce sprang up. Goods which had previously taken days to deliver could now be conveyed in just a few hours. Communication between markets also improved significantly. Produce and dairy products were in great demand in Chicago and Milwaukee for use locally and nationally. As the Civil War loomed, the northern states were at an advantage because of their transportation and industry.

Cuba Township. Built around the 1870s, the farmstead was arranged in an orderly pattern which included a milk cooling shed. The milk house was spring fed from an underground spring that allowed cold water to fill the concrete baths, where the milk would remain cold before it was taken to market. Almost all dairy farms had this milk cooling shed either directly attached to the barn or situated close to the front where the cows were milked.

In Lake County, there were many supporters of the Civil War. There is a story about the recruitment of young soldiers from Lake Zurich: "One farming family was working the fields in the morning and had taken a break for lunch. One of the sons placed his scythe in the big horse chestnut tree on the front lawn and saw that there was a group of soldiers that had come to recruit for the war. The son jumped on the wagon and headed off, away from his family to fight for the Union Army. The boy's mother was going to remove the scythe from the tree, but the father stopped her. 'He will remove it himself, when he returns.' In 1940 the tree was still standing, and the blades of the scythe were still protruding only a few inches from the crook of the big tree."

The settlers from New York and Pennsylvania were avid supporters of the anti-slavery movement. There is strong evidence of an underground railroad connection in the county, and it was not until late in the war years that a draft was needed to continue recruitment of soldiers. Incentives were invoked to recruit more men as the war raged on through the early 1860s. Young men who enlisted in August of 1862 were promised that they could stay away from training camp until the summer harvest of crops was completed on September 1. From 1861–1865 more than 2,000 men from Lake County joined the war in battle all across the various campaigns. Some reports claim a higher number because of the amount of men who registered in Cook County across the southern border and were then counted in the Cook County total. In a county of less than 19,000, this was an enormous percentage of the work force. Of these 2,000, more than 400 never returned. The war was felt to be a necessary evil to abolish the blight of slavery in the country.

The end of the Civil War in 1865 marked a period of relief for many farmers. Business returned in full force and Chicago was demanding the goods produced on the rural farms. Additional rail lines through the county were completed in the 1870s. The Wisconsin Central through Lake Villa was established in 1872, the Elgin Joliet and Eastern in 1889, and the Chicago, Milwaukee & St. Paul through Fox Lake in 1893.

The infamous Chicago fire in 1871 had a significant impact on the outlying areas. Trade was reduced temporarily with Chicago, and children orphaned from the fire were brought to the country by Catholic charities to find homes. The Weidner family of Buffalo Grove was one family that adopted a young daughter. Family members recall the stories about the wagons coming through town with the children, asking for families to take them. Some families needed the extra hands to help out and chose boys, who became loyal and active contributors to the farm's work needs.

Lake County was discovering that the land, while good for crops, was superior for dairy cows and beef cattle. Many farms abandoned other pursuits in favor of cattle. Between 1860 and 1870, the number

Cuba Township. The earliest sections of the barn date back to the early settlement period. Later sections were probably built in the 1880s and then expanded in the early 1900s. The soil banking probably occurred around the 1880 expansion. The farm was located according to traditional settlement needs—near water, woodland, and cropland.

Opposite page: *Ela Township. An example of second generation barns that were built larger. The barn was soil banked on a piece of the property accessible to the fields and with proper drainage in mind. Near the barn is the small milk cooling house indicative of a dairy farm.*

of milk-producing cows in Lake County increased by 41 percent as farmers expanded their herds by importing new livestock. The stockyards in Chicago grew to great size and there were farms near rail lines specifically designed for quarantine purposes, to hold cattle until they were cleared for health. One barn in Libertyville has been modified as a home, but was originally a quarantine barn. The lower basement has walls of stone twenty-six inches thick. Nearly every community along the rail lines counted on a creamery or ice house to store dairy products until they could be sent to Elgin, Joliet, Chicago, Milwaukee, Minnesota, and towns in Indiana. By the end of the 1800s, farming as industry in Lake County was firmly established.

SECOND GENERATION BARNS

In 1865, the population in Lake County had increased dramatically to 18,660 from the 2,905 of the first census in 1840. There were, by 1861, fifteen major settlements in the county: Waukegan, Forksville (Volo), Deerfield, Barrington, Diamond Lake, O'Plain Bridge (Gurnee), Lake Forest, Hainesville, Long Grove, Lake Zurich, Libertyville, Antioch, Half Day, Millburn, and Wauconda. Each boasted its own general store, blacksmith, wagon maker, tavern, church, and creamery.

As the county grew, these residents merged cultures and designed barns with advances from around the globe. The farmers constructing barns during this time built them bigger and more functional than previous structures. They were no longer exclusively for animals or just for wheat; they were now designed for multiuse. The German custom of soil banking was combined with the three bay floor design of the English builders. The larger barns were sometimes referred to as "prairie castles" for their great size. These barns were more important than the owners' homes and their size and care reflected this. Rick Bott, a contemporary barn specialist in Lake County, explained that when a farmer's barn was extremely large and the home very modest, it was felt that the farmer "truly had his priorities in order."

Ela Township. The Clark family farm, a good example of a "prairie castle."

Influenced by the Victorian era, farms were laid out in a square pattern with solid buildings that could be easily rearranged and moved. This allowed for neat and tidy site plans. Publications gave tips on this arrangement of farm buildings and on building farmsteads with the future growth of the farm in mind. Donald Berg, in *American Country Building Design*, stated, "Barns built with tight, mortised and pegged frames were incredibly flexible. Farmers hitched yokes of oxen and dragged them to new locations as needed. A homesteader's first barn could become a carriage house or the structure of a bigger home as the family prospered."

Below: *Ela Township. This farmstead built in the late 1860s is a perfect example of the neat and tidy farm arrangement of the Victorian period. The barns closely resemble many New England farmsteads. This barn was one of several in Ela Township built around 1862. Owned by the Vehe family for four generations, this barn has been preserved by the village of Deerpark and the Vehe Farm Foundation.*

The hardwood virgin timber that had been so readily available to the pioneers became scarce, not only because of local use, but also due to demand from other markets around the country. Lake County shipped timber from both the waterways and railways. As our timber left, other materials for barn building were coming into the state via the railroads. Wisconsin pine was delivered in enormous quantities for siding and even some of the main supports.

The shrinking wood supply forced farmers to seek alternatives to the large timber-frame construction. The construction and design of barns was still largely done by skilled craftsmen overseeing the preparation of the joints and the farmers and farmhands finishing the job. Ideas were based on neighbors' successes and rumors rather than engineering expertise. By the late 1800s new ideas were generated from the main farmbelt states and being slowly embraced by other areas. Improved milling of lumber provided less expensive planks for building. Plank framing, using smaller boards that relied upon distributing the weight of construction throughout the entire structure, was one such innovation.

Ela Township. A portion of the foundation of this barn was built with limestone quarried from nearby mines. The last family to own this farm had ten children who enjoyed the benefits of raising livestock. This was long after farming was a way of life in Lake County, when most other children in the community lived in subdivisions. Probably built around 1875, the age is indicated by the ridge pole on the roof peak that was added to accommodate the new hay carriers of the day.

From the agricultural schools and publications came suggestions from farmers all around the Midwestern states. One of these suggestions was the cupola. The cupola allowed for improved ventilation in the hay lofts. When hay does not dry properly, there is a chemical reaction that creates heat and may spark, causing spontaneous combustion of the hay. This was the reason for many unexplained barn fires. Before the realization of the spontaneous combustion, it was thought that the fires were caused by lightning strikes. Some farmers believed that this was God's will and therefore should not be deterred by the use of lightning rods, but the cupola solved this problem, while not falling into the category of being a "heathen" practice.

One of the most important inventions of the day was the hayfork, around 1860. This reduced the backbreaking work of men using a pitchfork to lift hay over their heads for storage in the lofts. When the subsequent hay carrier was invented by William Louden of Fairfield,

Iowa in 1867, roof design was completely transformed. The hay carrier required a rack built along the roof peak, or ridgepole, to carry the hay horizontally along the length of the interior roof line. An unimpeded roof peak was necessary to make the work of the hay carrier more efficient, saving the work of the farmer tenfold. By the 1870s and 1880s, the hay carrier was commonplace in barns, and roofs were higher, longer, and free of crossbeams in the lofts. The European influence of the gambrel roof (named after the hind leg or hock of the horse) further improved loft storage capacity and could be built using new roof designs with the hay carrier in place.

During the late 1800s, the popularity of exterior protection for the barn's surface increased. Some farmers strongly believed that barn siding properly cured would not need protection from paints or oils. Others relied on the neatness and beauty of paints. Early finishes to protect the exterior were made on the farm combining linseed oil (from flax), skim milk, and iron oxide from the soil (the iron oxide gave the paint the traditional red color).

By the end of the century, life in Lake County remained vastly rural. There were about 2,300 farms in the county, most producing dairy products. The many changes that occurred from 1850–1900 helped the farmers improve as businessmen during this time of great prosperity.

Ela Township. This barn was built by the Volling family around 1877. They moved the barn from its original location to its current spot. Another barn addition was added by the Clark family who owned several other farms in the area. The history of this barn has been preserved in the oral history archives of the Lake County Discovery Museum.

Ela Township. One of few farms remaining that still have an old windmill over the original well. The size and open design of the interior posts would place the date of the barn building to around the mid to late 1800s. Once a thriving family dairy farm, this farm now breeds horses.

Right: *Ela Township. Family-owned for more than one hundred years and one of the last working farms in the township, this family continued to raise dairy cows and beef cattle until 2000, despite the encroachment of subdivisions. The owner bought his feed from the same company in Crystal Lake where his family had done business for four generations.*

Ela Township. High on the best-drained part of the land, this barn sits on one of the original farmsteads in the township. The gambrel roof dates this barn to the late 1800s, but other barns on the property were built earlier. During the early 1900s, Arabian horses were raised here.

Opposite page, top: *Ela Township. One of the few barns that gives the date of its creation. This barn was built by the Schwermann family in 1877. It was built with the multicultural traditions of the day, incorporating the German soil banking with the English three-bay floor design. The interior was created using old hand-hewn beams and milled lumber. The white addition was built in later years to accommodate horses.*

Opposite page, bottom: *Ela Township. A perfect example of the prairie castle, grandiose in size. The air vents were a later addition to permit ventilation of the upper lofts. The cross was added to adorn the barn and remind those traveling by that the owners were a good Christian family.*

Ela Township. Built near the turn of the century. Family-owned for several generations, another tidy arrangement of buildings. In later years an antique business made its home here.

Ela Township. Several generations of families added onto this barn as evidenced by the poured concrete foundation and the cinder block foundation. Interior framework suggests construction around the late 1800s.

Facing page, inset: *Ela Township. In the upper lofts of the Miller barn a star was carved near the roof peak. Carvings of this sort were common on German barns on the east coast, but this is the only one found in this county. The carving was included to remind the farmer of his religious background, as well as to allow for a little light into the upper lofts.*

Ela Township. Originally built by the Miller family, this barn was once located farther west on the property. In 1930, new owners moved the barn eastward, placing their house on the original location. Horse stalls were also added in the 1930s. Lilacs were frequently planted around the doors to the barns and the outhouses, so that when the spring thaw arrived there would be something fragrant to counteract the typical barnyard odors.

Fremont Township. A large family-owned dairy farm. The family had several farms in the Fremont Township area. Built near the turn of the last century, this style was popular for large dairy businesses. The lowest level would have had a grand milking parlor to accommodate the large herd.

Fremont Township. This barn shows how new generations added onto the original structures as business grew. The final result of this barn initially built in the late 1800s was the T-shape design. Most notably, this farm was owned by the actor Marlon Brando, who purchased it as a gift for his aunt.

Above: *Fremont Township. Gabled roof, soil banking, and few windows combined the traditions of the day in this barn built around 1893. The hex symbol near the roof peak was a gift from a friend to the current owners. Hex symbols were common among the Pennsylvania Dutch communities. There is some debate whether they symbolized the wish to hex away evil or whether they were just decorations of pride and "Chust Pretty."*

Fremont Township. A larger barn on a property that contained the original village hall of Fort Hill and a dance hall built in the 1840s. This barn represents the largest of the century farms built toward the latter part of the 1800s. A true "prairie castle."

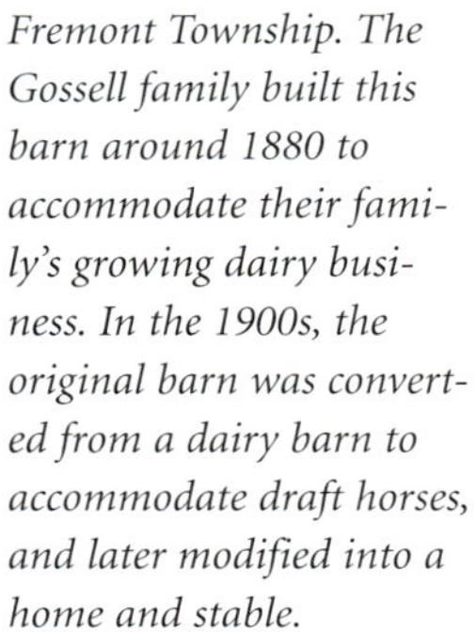

Fremont Township. The Gossell family built this barn around 1880 to accommodate their family's growing dairy business. In the 1900s, the original barn was converted from a dairy barn to accommodate draft horses, and later modified into a home and stable.

Fremont Township. Family owned for more than one hundred years, this barn was the property of the Holland family. The dairy remained in operation well into the 20th century.

Grant Township. This barn was built for a large family dairy. The ventilation windows over the doorway, the large threshing doors, and the multi-use structure were classic second generation barn features. As modern developments encroach, the family still maintains its old traditions.

Grant Township. Today the Volo Bog Preserve boasts acres of rare bog land, but the rehabilitated dairy barn had interesting owners over the years. At one time it was surrounded by the Wing and Fin club and the University of Illinois owned it. A gentleman commonly referred to as the "Captain" was the man who sold it to the Illinois State Department of Conservation in 1960.

Above, left, and opposite page top: *Fremont Township. For twenty-five years the Archdiocese of Chicago owned several hundred acres of land purchased for its potential use as a cemetery. The land was rented to farmers in the interim until its use could be decided. All three barns are synonymous with styles of the mid- to late 1800s. Soon an industrial park will replace these farmsteads.*

Grant Township. The original property owner was Levi Waite, one of the early settlers of the area. The north-south section was built in 1889 while the east-west section was added in 1907. During the early 1900s, Mr. Frost owned the farm, thus the name on the hay door. The gable roofs included roof peak extensions, or "rain hoods," to protect the hay while it was raised into the lofts.

Grant Township. A horse training center resembling the styles of the mid-1800s. The tiled silo dates from the 1920s when various materials for construction of silos were being tested.

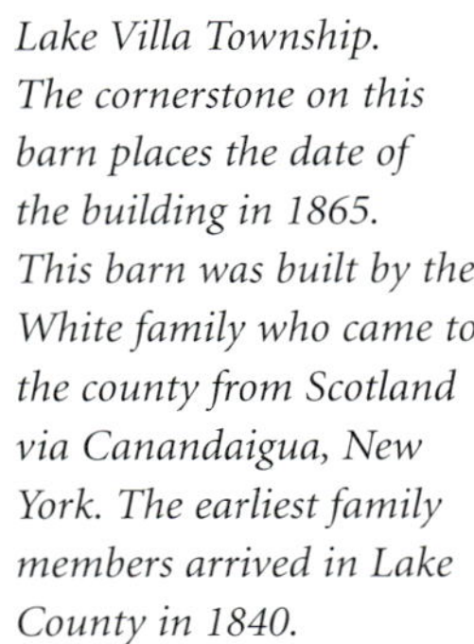

Lake Villa Township. The cornerstone on this barn places the date of the building in 1865. This barn was built by the White family who came to the county from Scotland via Canandaigua, New York. The earliest family members arrived in Lake County in 1840.

Lake Villa Township. In the tradition of the second generation, this multi-use barn was large in size, still possessing the gable roof and few windows typical of the early generations. The silo built around the turn of the century indicates a dairy farm.

Grant Township. While the barn dates back to the late 1800s, the rafters in the roof of this barn were added in the early 1900s. The same carpenter also replaced the Deidrich roof. (See Chapter 4)

Libertyville Township. A smaller dairy barn owned by the Hoffman family in 1873. The roof line is indicative of the European influence of the gambrel. In the 1950s the barn was modified to accommodate horses.

Libertyville. The Brogden family came from England and purchased this farm in 1864. They bought 105 acres for $1,500. The barn was built in 1865. Initially sheep were raised by the family until the dairy business became a more desirable industry. The cooling shed is spring fed. The tilted window near the roof was installed at this angle to allow it to be placed as close as possible to the roof peak.

Lake Villa Township. This barn contains all the features of the second generation farms built in the mid- to late-1800s. The neat and tidy arrangement of buildings, the multiuse design, and window or light panes above the door to the three-bay barn were all second generation features. These windows were often louvered so that they would allow light as well as air into the central floor. The barn also has a rain peak at the roof line of the loft and a traditional cupola.

Newport Township. Once a thriving dairy, this arrangement of buildings dates the farm design to the mid- to late 1800s. The brick silo would have been built around World War I, while the concrete silo would have dated to around the turn of the century.

Newport Township. This farm was once owned by a Bohemian settler from Chicago named Otto Maria. Otto served as the Newport Township assessor. The farm now operates as an independent lumber mill.

Newport Township. Owned by the same family since 1859 and registered as a Centennial Farm in 1978, this farm was once part of the Corris brothers' dairy.

Newport Township. The original section of this barn was constructed around 1860. Additions were made through the years as the business expanded. Current owners raise 200 varieties of chickens.

Newport Township. Originally the land and farm were built in 1865. William Strahan, who owned the property, kept the barn and outbuildings well maintained in the Victorian manner. On the property is an old circular saw that could have been powered by a steam engine and used by the local neighbors. Circular saw designs are credited to Shaker farmers around the mid-1800s.

Newport Township. Once owned by the Welch family and built around the late 1800s, this large dairy barn was eventually sold to the many holdings of Tempel Farms who presently own several old homesteads in the Wadsworth area.

Vernon Township. Family-owned since 1912, the barn dates to 1892.

Shields Township. This barn, later converted to a business, was owned by one of the first African American residents of Lake Forest who ran a livery service. Built in the late 1800s, the barn was part of the servants' section of town in Lake Forest, now an area of residential housing.

Vernon, West Deerfield Townships. The old Conway farm. Miles Conway came to Lake County around 1864 with eleven children and settled on 116 acres. Each subsequent generation of family added to the length of this barn. By 1999, this was the last farm in the area to still raise black angus cattle, known for their quality beef.

Sketch illustrates combining features of an English wheat barn with a German bank barn for a multiple-use building.

Vernon Township. Frank Hoxie built this barn in the late 1800s. While altered for use, the original arrangement of buildings has remained intact on one of the last surviving farms in Vernon Township.

Vernon Township. Encircled by industry, this last remnant of farm still maintains its Victorian pattern. The farm was purchased by the current family in 1893 from the original land grant owners.

Warren Township. A classic second generation farm. The silo top is one of the early cone designs. Later silos were topped with domes.

Warren Township. A "prairie castle" with gabled roof. Many barns remain intact for storage and crop farming, even as dairy farming has been centralized to large facilities.

Warren Township. The original section of this barn was built in 1875 by members of the Bonner family. The next generation provided the addition, creating a T-shaped barn. This style of adding on to the barn is common in the Midwest, especially in northern Indiana and Ohio.

Wauconda Township. The same family has owned this barn through four generations. The first family member was recorded as having come to the area in 1858. Many additions have been made to the original barn over the generations. Some were added on the exterior without any change to the interior walls. This family has always raised Jersey cows, most noted for the richest milk in percentage of butterfat and protein.

Wauconda Township. A late 1800s barn with a new horse training center attached. The farm dates to 1849, but the barn's roof line indicates a later style.

Warren Township. Another large, old classic. The barn's size would accommodate a large herd of dairy cows. The soil banking on this barn was on the short side of the barn, unusual for the time, but this allowed the cows to be milked in the lower parlor and then led to the fields to the east of the property.

Wauconda Township. A T-shaped barn. The buildings here were attached in a continuous line. This form of "continuous" architecture was popular in New England to protect the farmer from the elements during inclement weather.

Wauconda Township. The Robertson farm, part of the Lake County Forest Preserve. This grand dairy farm combined the German soil banking, the three-bay English design, and the European gambrel roof to form a large multi-use "prairie castle."

3

THE GENTLEMEN

BY 1890, CHICAGO WAS BACK ON ITS FEET FROM THE destruction caused by the great Chicago fire of 1871. Wealthy industrialists became bored with city living, and the country was calling to them. In search of a slower and more rural lifestyle, the gentlemen farms of Lake County were born. Sometimes these owners were called checkbook farmers because they came, wrote a check for the land, and moved in. There were three distinct cycles of the gentlemen farms.

The first cycle came around the early 1890s. These were the first industrialists who were building their country homes in proximity to their country clubs. Games like polo were popular with this set, and the surrounding land was needed to house the polo ponies. The farms were run and managed by others; the country gentlemen were simply visitors, and there was no motive for profit on the farm. The estate farms were designed and built by leading architects of the time. Some claim that the industrialists built the railroads to the suburbs simply to have easier access to their country homes.

The second surge of gentlemen came between 1904 and 1929, when the first automobile came to Lake County, which made accessing the land beyond the railroad much easier. The estates were the epitome of the perfect farm. Aesthetically and functionally pleasing, the estate grounds were designed by landscape architects. The animals were cared for like family, and some of the gentlemen during this period enjoyed the farm so much that they made it their main home. The farms were designed with the most modern accoutrements of the day. Hot running water, central heating, and electricity, all luxuries still several years away from most country dwellers in the early 1900s, were commonplace on the gentlemen farms.

For many country farmers living in the area, these new farmers were a curious lot. Some farmers, whose families were changing and leaving the business with this next generation, sold the land at a profit and continued to work the farm. A farm manager's family enjoyed

Libertyville Township. Built around 1920 by Herman Helfer, this estate was known until the 1960s as a showplace for prize-winning dairy cows and owned by Gaylord Donnelly, one of the large land owners in Lake County during the mid-1900s. The estate has been maintained and renovated in recent years.

farming without the responsibility of ownership. Other farmers rented land from the new gentlemen on a share-lease program, where the farmer produced and shared the profit with the owner, in exchange for use of the land.

The presence of these gentlemen farmers enhanced the life and businesses of the local farmers. The modern convenience of the railroad and the quest for road improvements helped farmers get their products to other markets with greater ease, and some gentlemen expanded the farm to include a dairy, which sold the products of the local farmers. This brought the businesses closer to the locals, again without the responsibility of ownership. The advantage that the gentlemen had over the smaller farms was that they could afford to maintain their businesses according to modern standards. Another advantage was that, being businessmen by trade, these gentlemen saw the need for home delivery in neighborhoods where other country gentlemen and estate owners lived. They began the competition that later affected many farmers in the quest for adapting to the changing times and demands of commerce. (Chapter 6)

Cuba Township. The Von Haegen family estate was originally called "Wakefield" after the family's original plantation in Virginia. The Von Haegens were prominent investors in the steel and railroad industry in Chicago. The red barn, built in 1907, measures 180 feet after an addition to the barn was made in 1908 to the existing 120 feet. Originally, Percheron horses were raised on the farm. Today, the structure is home to a few horses, many cats, a law office, an apartment, and a great deal of storage space.

After the stock market crash of 1929, the third cycle of gentlemen farmers came into existence. Many of the earlier gentlemen farmers were forced to sell off their land or split the acreage. In their place came a new crop of gentlemen, some the sons of these same industrialists, who again took advantage of the land prices and purchased many small farms. This was a period of suburban growth—the first of the city commuters. Roads were beginning to be paved, and these commuters could now live on the farm and work in the city.

One humorous story from this time period involved a family in Barrington. The "new" farmer wanted fresh milk on his farm. Consequently, he hired a farm girl and bought a cow that he presented to her for approval. "Well," she said, "She is a real nice cow. I know a farmer nearby who has a great bull we can use." The gentleman queried, "Why do we need the bull?" It took a time before he understood the spasms of laughter that the girl broke into after this question.

Cuba Township. The calving barn on the Von Haegen estate. The barn measures 120 feet in length. The addition to the apartment in the loft was made in 1998 by local architect Robert Parker Coffin. The original conversion from dairy barn to living complex was completed by Charles Pope in the 1970s. When the barn was used by the Von Haegens in 1907, the calving cows were treated to hot running water, electricity, central heating, and cork floors, so their toes would not get too cold in the winter. The original estate had eighteen farm buildings.

The 1930s were marked by a renaissance of construction for the gentlemen, some building grandiose dairy showplaces and some having the earlier barns modified by new architects to accommodate new trends like fox hunting and horse training.

Several of the estates of the early gentlemen have withstood the test of time. In part, this endurance came from the buildings being so well built and well maintained. Even as the main estate homes were torn down because of their massive size during the subdivision of large estates, the carriage houses of the estates, where the horses lived, were often saved and converted to homes. These carriage houses dot the North Shore and many other communities throughout Lake County.

Times have not completely lost the gentlemen farmers. Some may say that there is a fourth generation today. There will always be those of affluence who prefer to live in country homes, and many do appreciate the old farm buildings, or still use them for their horses. There will always be those who long to find a simpler, more genteel lifestyle in the country, closer to the earth and nature.

Cuba Township. In the 1930s, many farms were purchased throughout the Barrington area as weekend retreat farms. Some of the owners fell in love with the farms and moved there permanently. This farm was owned by the Leonard family, who would host fox hunts between the neighboring farms.

Deerfield Township. This pony barn was built in 1894 by Howard VanDoren Shaw, a prominent architect on the North Shore who is most noted for the design of Market Square in Lake Forest. Here, the owner raised polo ponies. Most remarkably, the farm is remembered because F. Scott Fitzgerald modeled his character, Daisy Buchanan of The Great Gatsby, *after Ginevra King, the daughter of the estate owner. In his book, Fitzgerald mentions the farm in passing when he says, "Tom Buchanan brought down his polo ponies from Lake Forest."*

Ela Township. Once owned by the Wilke family who owned Do-All Industries in Des Plaines. Willowbrook farm was a functioning dairy farm on 160 acres. The main barn also housed an art studio and dance studio at one time. The homes were surrounded by limestone lagoons and pools, and the main home was considered to be quite elegant for the time.

Fremont Township. Chardon Farms was named after Charlie and Donna, the children of Ray Krop, owner of Krop Forage. Each of the barns on the property still has the name "Chardon Farms" carved above the barn doors and etched in the cement of the foundations.

Fremont Township. The yellow barn on Chardon Farms was built in the 1930s, and it is still used for domestic animals today. Many of the barns were designed with the "prairie style" roof line with extended bays. This roof line was felt to deflect the sometimes harsh winds of the prairie.

Lake Villa Township. When this Lehman family dairy barn was transferred to the Episcopal Diocese in the early 1950s, the barn still had dirt floors and hay in the lofts. During the conversion of the barn to Holy Family Episcopal Church, the doors were sealed where the original creamery entrance was located. The original central loft area is now appropriately the church sanctuary: There is old saying that a farm boy's sanctuary was the lofts of the barn.

Libertyville Township. Built around 1950, this estate farm hosted sheep, horses, and dogs. For the owners, the estate was a beautiful retreat from the city, with picturesque views from the home that overlooked the expanse of rolling property.

Libertyville Township. Originally Samuel Insull built his Italianate mansion and estate on 4,400 acres in 1911. The grounds were designed by landscape architect Jens Jensen. In 1929, when Samuel Insull lost everything in the stock market crash, the estate was sold to John Cuneo, a publishing giant. Cuneo took the farms to new heights, producing Hawthorn Melody dairy products and creating a farm where children, particularly the underprivileged, could come and view the farm life up close. Hawthorn Melody farms boasted one of the few round barns in the county, exotic animal exhibits, and a milking parlor with glass windows so that observers could watch the process. The last remaining barn on the farm is this one. Once exotic animals were housed here. In later years the barn was the home and garage for Mrs. Cuneo's driver, and today the barn is the residence of the curator of the Cuneo Estate and Gardens.

Libertyville Township. This barn, originally built in the late 1800s, was modified during the 1930s by Stanley Anderson, an architect known for his conversion of dairy farms and his distinct style of farm buildings. Anderson updated the dairy facilities and added stables for horses.

Libertyville Township. The last remaining carriage houses of architect David Adler's home. These barn buildings were distinctive for the architectural features favored by Adler. Many of his estate farms still exist in Lake County today. The buildings are now home of the Adler Cultural Center, where art and music continue to flourish.

Shields Township. Built to resemble a Normandy dairy farm, this estate farm was begun in 1923 by Clifford Leonard. The project was abandoned after seven buildings were completed—the chicken house, three barns, and three cottages. The proposed Normandy chateau was never built. After being abandoned for twenty years, the estate was updated and transformed into luxury homes. The large cattle barn was converted by David Barrow in 1958. It boasts 140-foot ceilings in the central living area, with 26-inch-thick walls and an underground bomb shelter in the converted silo. This building is the horse barn.

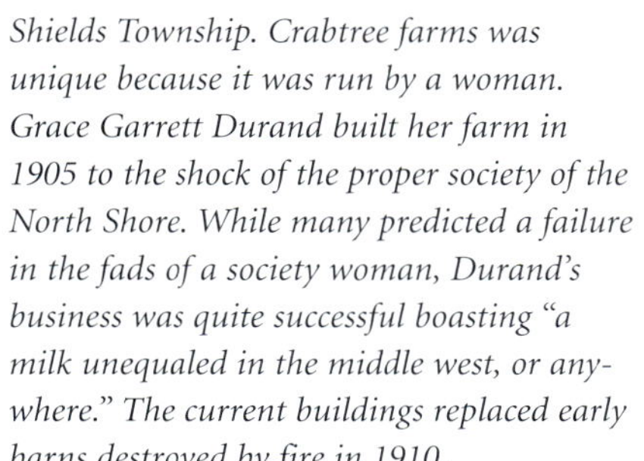

Shields Township. Crabtree farms was unique because it was run by a woman. Grace Garrett Durand built her farm in 1905 to the shock of the proper society of the North Shore. While many predicted a failure in the fads of a society woman, Durand's business was quite successful boasting "a milk unequaled in the middle west, or anywhere." The current buildings replaced early barns destroyed by fire in 1910.

Shields Township. A. Watson Armour built this country estate in 1917 so that his family could visit the countryside on the weekends and play farmer. The original design was created by Alfred Hopkins who styled farm groups in the Georgian Colonial Revival style. The farm was named Elawa after Elsa and A. Watson. One of Hopkins' design features was to separate the animals to avoid contamination, while connecting the buildings so that the owners could move from one stable to another without being exposed to the elements.

Vernon Township. Built in the 1930s by the Ryerson family, this estate was called Brushwood Farm. The Ryersons raised Arabian horses here. In 1966, 257 acres were donated to the Lake County Forest Preserve, and today Ryerson Woods boasts 550 acres along the Des Plaines River. The old barns still host animals that are examples of rural farm life for area families.

Wauconda Township. Lakewood Farms. Malcolm Boyle, a general contractor from Chicago, purchased 1,250 acres of land in the Wauconda area around 1937. He had lagoons and lawns designed by local landscape designer Synnestvedt and his buildings were designed for beauty rather than mere practicality. The farm boasted show-winning Guernsey cows and played host to local livestock functions.

Top photo: *Warren Township. Owned by the Bartholomay family and remodeled by Stanley Anderson, these horse barns are connected in the New England-style of continuous architecture. This style was adopted to protect the family tending to the animals from the elements during inclement weather. The added architectural features of fancy cupolas and dormers were common during renovation of old farms in the 1930s and 1940s.*

4

1900–1930

IT WAS THE TURN OF THE CENTURY AND MODERNISM WAS the order of the day. In 1900, Lake County boasted 2,229 farms primarily producing dairy derivatives for sale in Chicago, Milwaukee, Elgin, Joliet, and other parts of the country serviced by these cities. The Midwest Industrial Revolution, in full swing, was influencing the way of life for farmers as well as merchants. Also impacting the quality of life on the farm were world events and health issues for animals and humans alike. These challenges were some of the reasons that there was a reduction in Lake County farms by the end of 1930.

Industry influenced farming in several ways, mostly causing a shortage of farm labor. Many farm hands were lured to the city to work for assembly lines. The pay was better and the work less labor-intensive. As they matured, farm children, with their schooling, were also attracted to the city, where education was needed in business. In years past they had applied their education to the farm. On the other hand, products of industry improved the efficiency of farm machines and required less manpower. Farmers, always friendly with their neighbors, formed cooperatives to purchase new machinery and share it among themselves.

Tenant farming was also an effort to maximize the workforce by sharing the land, machinery, and products of the farm. Tenant farming allowed the landowner to rent the land for lower prices to the farmer in exchange for a share of the crops and production of farm products. By the 1920s there were approximately 800 tenant farms in the county that were included in agricultural surveys.

World War I created a frenzy of support for the war effort. With the shortage of supplies, involvement in the war effort was a unifying but trying period for farmers and businesses. Just as Lake County had contributed so much during the Civil War, this war also attracted a large portion of the 21- to 31-year-old male population of the county. When asked to produce more wheat for the war effort, Lake County obliged by contributing 1.8 million bushels, which was 1.2 million

Ela Township. On property once owned by George Ela, recognized as the settler of Ela Township, this early 1900s barn is similar in style and construction to the mink farm across the street, showing how neighborhood trends were common. One unusual feature on this barn is the rooster design imprinted in the concrete foundation.

more than the initial goal. The population at the time numbered around 55,000, and it is estimated that 4,200 served in the armed forces during the war.

Antioch Township. The owner of this barn ran a pub in downtown Antioch. When Prohibition was declared in 1920, he was forced out of business and went into dairy farming. The barn was built shortly after 1920.

Aside from the interruption of the war, health problems increased awareness of a need for sanitation standards on the farm. In 1908, consumption created an epidemic in rural areas. Around the same time, anthrax wiped out several farms' livestock, and hoof and mouth disease, if contracted, meant quarantine and the destruction of the entire herd. Samuel Insull's herd and a herd of a nearby neighbor contracted this disease and their cattle were driven into a large trench, shot, and spread with quick-lime in an effort to eradicate the plague.

Livestock were not the only victims of illness. The influenza outbreak of 1918 took thousands of human lives all over the country. Schools were closed due to the widespread epidemic that is estimated to have affected more than 8,000 people in a one-month period in Lake County alone.

Another health problem that affected many farms was the tuberculosis epidemic of 1926. In her diary, Irene Mills Snetsinger, who had a farm in Lake Zurich, documented its effects. "The county veterinarian came to inject the cows with vaccine for the TB tests in April. Three days later, when the tests were read, we had three cows and three heifers left out of a herd of 27 and no money to buy replacements. The cows that didn't pass the test were branded TB and shipped to the stockyards for tankage or fertilizer. We were to get a cent a pound for them from the government, but this payment didn't come through until September. All of the herd in the county were tested at this time. We didn't have any income that summer. We had a garden and canned vegetables and fruit. . . . Finally, the check from the government came for the cows in September. As it was only for one cent a pound, it was nowhere what the cows were worth as dairy animals. The barn had been disinfected and we began to buy cows to replace those lost in the test. Since all of the farmers in the county had lost cattle, replacements were hard to find and much higher in price. From then on, the veterinarian came twice a year to inject the vaccine and we would lose more cattle. When he came, I would start the fire to heat the branding iron that he used to mark TB on the cows and then I would cry. . . . After about four years, we had a test and no reactors were found and we began to get our herd built up again."

For families living along Lake County waterways in the early 1900s, the potential for income from other sources was too much to resist. During summer months, city dwellers would flock to the numerous lakes and rivers, transforming sleepy farming communities into summer resort towns. Homes were converted to "bed and breakfasts" and cow pastures were used as picnic groves. Any farm that had been built next to a lake was cleared for recreational purposes during this time of resorts. The resort business remained active until the onset of World War II.

Problems resulted from these resorts during Prohibition (1920–1933), because unsavory characters felt that Lake County was a place where illegal activities could be carried out without close observation from the law. It is rumored that there was a safe house in Villa Park near Deep Lake and Fourth Lake where mobsters could meet to negotiate business. Terri Druggen, Al Capone and "Baby Face" Nelson owned farms in the county. The farms caught in the crossfire were few, but the area was still affected by this era of crime.

During the winter, the same home that took on summer boarders would provide boarding for workers from nearby ice houses. The ice harvesting season, which attracted nearly 10,000 young workers from the city, began in December and ended in March during good years. Several ice manufacturers set up giant ice houses around the lakes of the county. Some of these companies would build off site and rent the lake from the owners. Local farmers would loan their work horses for the hoisting of ice, and farmhands could pick up extra work hauling ice.

In 1929 the stock market crashed. By 1930,

Antioch Township. An old carriage house in downtown Antioch, probably built around the turn of the last century. Carriage houses were common in more urban areas. The carriage house was a smaller structure that would hold the carriage, a couple of horses, and hay in the upper lofts. Many were built in a style that later converted easily into a garage for the automobile.

Avon Township. This family farm was owned by the Churchills. The older section, built around 1880, is gone completely, although clues to its existence remain in the discoloration of the walls of the new section, built around the late 1920s.

it was obvious that there had been a shift from farming to urbanization. Large farms were sold off to smaller farmers, and estate homes were subdivided into less extensive properties. Roadways and automobiles brought waves of commuters to the county for summer fun, and many imagined it would be nice to stay year-round. Suburban communities were growing along Lake Michigan and in pockets throughout the county. The first signs of organized developments were forming along roadways improved for auto travel. These new residents were urban commuters who were not interested in farming. They bought old farms cheap so they could live in a semi-rural setting, while envisioning ideas for improvements in education, shopping, and transportation.

By 1930, the number of farms in the county had dropped from 2,229 to approximately 1,566. There were still sleepy rural communities in the county, but their faces were changing as rapidly as industry was changing in America.

EARLY TWENTIETH CENTURY BARNS

Just as the Industrial Revolution had changed the way of industry, there was a building revolution in the business of barns. This was brought about by the shortage of large timber, by regulations requiring changes in barn space, and by the evolution of the farm. Barns built in the early part of the 1900s varied greatly in size, interiors, and roof designs from their predecessors. This was the age of experimentation.

Initially, in the early 1900s, barns were for the most part traditional; the large century farms of the late 1800s were still being built on a limited scale. The difference in these structures was the roofs. Many of the barns utilized the early timber-frame architecture combined with the large balloon frame gambrel roof. The balloon frame roof, as in the late 1800s, opened up the lofts for easy access and space. Resistance to other plank framing for the main supports was attributed to those farmers who believed in recycling the old wood of the early timber-frame barns. These farmers were the first to recognize the significance of the early timber. They were the original preservationists of agricultural structures.

However, preservation sometimes conflicted with practicality. The focus on dairy farming required a more consolidated farm. General stores and local businesses could now provide products from markets around the country that would simplify the farm family's life. No longer did the farmer have to produce everything himself. The concentration of the business end of farming was on dairy production and this, combined with the cooperative purchase of farming machinery by neighbors, meant that smaller, more efficient spaces were needed. These spaces would store feed and provide a sanitary milking parlor.

Dairies were required to meet sanitation standards that demanded newer, cleaner facilities if they wanted to remain in the dairy business. These standards included concrete flooring that could easily be cleaned, whitewashed walls, and many windows in the milking parlor for improved lighting and ventilation. Milking stalls were now required to be located at the end of the milking parlor within a short reach of the milk cooling bins, guaranteeing that the milk would be cooled quicker. With these requirements, it was often easier and more cost efficient to construct a smaller barn, under the new guidelines, rather than modify the current large barns to modern standards. These requirements and the overall consistency necessary in dairy operations created a high demand for smaller, low-cost, easy-to-build structures.

Avon Township. These two barns, located in close proximity, were probably constructed by the same builder, evidenced by the similar window designs and the roof pitch. This was a common practice, especially during the early 1900s, when local craftsmen dominated the market in certain neighborhoods.

The farmers' needs were answered by advances in plank framing methods. Improvements of light frame construction of barns were studied by engineers and agricultural stations around the country. The best plank framing methods utilized the Shawver truss design and the balloon frame roof design because they were easy to build and required relatively inexpensive materials. Rather than the experimental designs of the late 1800s, these barns were structurally intact. Too many early designs were based upon the idea of another farmer who perpetuated the design flaws of his neighbor who had used the same plan. Companies like Louden and Gorden, Montgomery Ward, and Sears Roebuck provided kits for building these balloon frame barns. The kits were welcomed because, in addition to the low cost, they came with easy-to-follow plans for construction, and building materials were delivered directly to the site.

One example of these kit designs, used in the early 1920s, taught the construction of the arch roof barn. These roofs were formed by wood cut into long strips and then glued together to create joists. The end result was unusual and aesthetically pleasing. The interior was like the inside of a barrel. This design can be seen only in select neighborhoods—it was more trendy than universally embraced. Some farmers criticized the style saying that it was "reflective of modern industry rather than the industry of the men."

By the end of the 1920s, few, if any, new barns were built with old wood traditions. Plank framing had been embraced for its convenience and cost. The loss of children to industry and the unionization of unskilled labor had its effect on traditional farming. Barns were for dairy farming or specialized family needs. From the exterior, a passerby could easily determine the business of the family who lived on the farmstead.

Benton Township. Built by the current owners' family in 1914. This pre–World War I barn resembles the early wheat barns constructed by the English settlers. Few windows and the simple design were all that was needed for a small family farm.

Avon Township. One of few barns remaining around the lakes region of the county, this structure functioned as a carriage house and garage in its lifetime rather than a livestock barn. The windows in the upper lofts, in particular, indicate a lack of routine hay storage

Deerfield Township. A barn owned by Irish immigrants who came to Lake County during the great potato famine. The original farm consisted of 200 acres. The family raised crops and livestock. This barn was built in the early 1900s after the original barn burned to the ground from a lightning strike. The unusual brick foundation and horizontal siding make this barn a unique design for Lake County.

Ela Township. Interior construction resembled barns of an earlier period, but this specialized barn was built in 1905 by the Peppers, a German family living in the Lake Zurich area. Hidden in the wall of the barn was a note in a citrate of magnesia bottle. The letter read: "Lake Zurich, Ill. Sept. 8th, 1905. This is time the wall was made of this barn. It is expected to be finished tomorrow. It was started Monday noon, Sept. 4th 1905. Mr. John Westfall was the masson bauss, Jack the stone bauss, Fritz the mud bauss, Louis the dirt bauss. The lumber is not here, it is expected any hour just as quick as it is here Mr. Ernst Branding the carpenter will start to work. Otto is the finishing bauss. Gust is the timber bauss. How many there are coming we do not now. The lumber cost $700 the masson work $50 the other expense is not knoed. Room for 36 cows and 50 ton of hay. 8 sedder post to keep the roof up the outside frame is 2x6 18 ft high with roof sidings. This is all about the barn. Their is 6 in our family, Grandpa, Mr. Henry Pepper Sr., Pa, Mr. Henry Pepper Jr. Ma, Mrs. Barrie Pepper, Son Louis H. Pepper, Daughter Bertha Pepper, Daughter Marie Pepper. Best regards to the one ever reads this."

Ela Township. A dairy farm of the early 1900s currently tucked away in suburban streets in spite of surrounding development. The barn was built during the resurgence of smaller dairy farms in the early 1900s. The farm still boasts the wind-powered water mill. The milk cooling shed, the beautiful barn, and the old farmhouse have all been preserved in excellent condition.

Ela Township. Constructed in the early 1900s by a builder from Palatine who owned a hardware store. The current family purchased the farm in 1950, and at one time they raised mink on this farm.

Ela Township. Near the town of Gilmer and located next to the rail line, this farm (built around 1925) was owned by one of the Umbdenstock family who were prominent in the Ela Township area. Proximity to the railroad meant that products from nearby farms would be delivered to markets more easily than those from farms located farther away. Gilmer was a stop along the Elgin, Joliet, and Eastern Rail Line.

Fremont Township. The arch roof evolved just prior to World War I as an alternative design, using planks instead of beams. The farm is typical of a smaller dairy operation. Hay filled the lofts and windows lined the milking area in the rear of the barn. The silo was positioned at the end of the feeding alley for easier access to the animals. The covered roof leading to the silo allowed for protection from the elements.

Fremont Township. An arch roof barn in Fremont Township. The roof style was experimental and frowned upon by some who felt that it served as "an example of modern industry, not the industry of man." Even the milk cooling shed had the arch roof.

Fremont Township. Singing Hills Farms. A large dairy operation owned by the Neilson brothers who ran a business in the early to mid-1900s. The barns varied from the earlier gambrel roof design to the more modern arch roofs of the 1920s. The arch roof was a popular style around World War I when plank framing was a necessity because of wood shortages. The large milking parlor of the main barn was needed for an operation of this size.

Fremont Township. A small horse barn built around 1915 for a Mundelein family that resided in town. The barn functioned as a carriage house and stable and is one of the last remaining barns in the downtown area.

Fremont Township. Built around 1913. The original owner immigrated to Lake Zurich from Germany when he was sixteen years old. He married and purchased this 107-acre farm in Mundelein where he ran a dairy business. Other family members settled on neighboring farms.

Fremont Township. This family-owned farm continues to function. The original Diebold family members came to the area around 1900 and settled on several farms. One hundred years later, there are several families that still share the neighborhood and carry on the farming tradition.

Fremont Township. The earliest record of the Hertel family living in Fremont Township is 1860. John Hertel served as the post-master for Fremont Center in 1879. The family currently owns several farms including this one, built around 1904.

Fremont Township. Originally built by the Holland family in 1902, this barn is the classic size and style of the early 1900s. The barn has been family-owned for the last seventy-seven years.

Fremont Township. The Ray Lake Farm specialized in Black Angus cattle, known for their quality beef. A large barn fire nearly destroyed the business and wiped out several older buildings.

Grant Township. Built around the turn of the last century, this farm was owned by the Deidrich family who owned several barns in the area. The older, more traditional style of this barn was favored by more conservative farmers. The family farmhouse and cemetery were located across the street from the barn.

Grant Township. Another Deidrich family barn. The original barn on this property burned to the ground during a July 4th celebration in 1921. The roof design is a newer balloon frame gambrel roof, similar to several others built in the area during this time period.

Grant Township. The concrete foundation, high-pitched roof line, and lack of threshing doors indicate the date of this barn's construction to be around the late 1920s. Other buildings on the property date to early settlement.

Grant Township. This barn was built in 1925 by Frank Heronymous who lost the property during the hard times of the Great Depression. During the 1930s and 1940s, Frank had a business whitewashing the interior of barns in the area.

Lake Villa Township. The cornerstone says the farm was built in 1901. Family members remember the beef cattle and horses that were family-raised. Also recalled were the many animals that strangers would drop off at the farm.

Libertyville Township. An arch roof style with soil banking. This dairy farm was tenant-farmed for several decades. The family raised goats, sheep, dairy cows, and many dogs throughout the years.

Libertyville Township. The framework of the interior places this barn's construction in the late 1800s, but so many modifications over the years and the use of the space put the real date of construction in the early 1900s. In 1924, WLS broadcast a show from this barn every Saturday from 12 p.m. to 12:30 p.m. called, "Man on the Farm." The Quaker Oats company owned the farm during this time as a chicken test farm. The farm was called the Full-O-Pep farm. The manager of the farm was a man named Homli Kent who was encouraged to move from his New England home to this area. When he resisted the invitation, Quaker Oats promised to reproduce his old New England home here board for board.

Libertyville Township. Built around 1915, this barn incorporated elements from the earliest barn on the property. The current owner has been planting black walnut trees for the last forty years for his grand-children. During the Columbian Exposition of 1893, there was a trench dug through the property and adjoining properties to bring artesian water from Kenosha to Chicago, where it could be enjoyed by patrons of the Exposition. After the Exposition, the canals were removed and the property returned to normal.

Newport Township. Originally the barn was built in 1929 for a family dairy business. The structure was rehabilitated in 1995.

Newport Township. The original building constructed around 1930 was collapsing, so the new owners poured a foundation, lifted the old framework, and set it upon the new foundation during the 1970s.

Newport Township. Wells Brothers Dairy, where registered Holsteins, known for the most milk production of all cows, were raised. Holsteins do end up as beef cows at the end of their lives. The barn was built in the early 1900s.

Newport Township. When a tornado blew the roof off this building in 1929, the Leable family replaced it with a gambrel roof and added a haymow. The farm was owned by the Ray family until the 1980s.

Shields Township. A small carriage and horse barn, this building has been immaculately maintained since its erection, probably the late 1920s or 1930s.

Vernon Township. The original barn built around 1900 was constructed for raising workhorses. The old carriage and wagons and the ribbons earned by the original owner were found in the barn when it was purchased by a new owner in 1969. The previous owner was the Lake County Deputy. The house still had the original Montgomery Ward sink fixtures and plumbing described in an old catalog found in the basement.

Vernon Township. Once owned by the Stancliff family and built around 1900. The small gambrel roof and foundation were restored in the 1970s. Today, children in a summer camp on the property enjoy the old barn.

Left: *Warren Township. Built along the O'Plaine Road, so named for the O'Plaine or Des Plaines River that flows nearby. This family farm, owned by the Thomas Rudd family, existed until the family members who farmed it died out. The barn was eventually sold to developers at the end of the 20th century.*

Below: *Warren Township. This barn was built in 1916 for $12,000. The barn frame was raised in one day, but a large windstorm blew it down. The second time it was raised, the ridge beam displayed a split where it had suffered damage in the storm. The 36-foot planks in the front are indicators of the length of wood available at that time. The original barn was intended for loose hay storage. It was owned by Olsen of Olsen Implements at one time.*

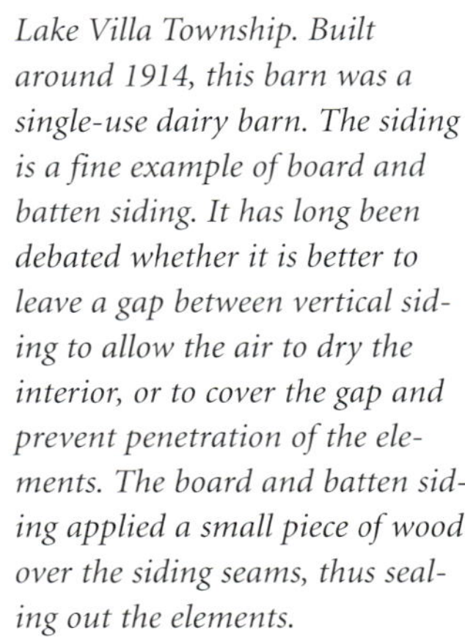

Lake Villa Township. Built around 1914, this barn was a single-use dairy barn. The siding is a fine example of board and batten siding. It has long been debated whether it is better to leave a gap between vertical siding to allow the air to dry the interior, or to cover the gap and prevent penetration of the elements. The board and batten siding applied a small piece of wood over the siding seams, thus sealing out the elements.

Warren Township. A replacement for the old pioneer barn that stood on the property, this structure is similar to several other arch roof barns built in the Wadsworth area around 1929. A family, who lived nearby, remembers the Saturday night barn dances held in the lofts that first year after construction, because the barn was built too late for the hay to be loaded into storage.

Wauconda Township. This barn (above) was built by the Walton Builders for the Henckel family to replace a log structure. Mr. Henckel received the land as payment for fighting in the Civil War. The second barn, across the road, (right) was also built around the same time using a Hendes Plank frame. In the early 1920s, a tornado blew through the area and moved the barn. The barn was pulled together with bridging and rebuilt with huge bolts. The barn was moved again in 1948 when they widened Route 12. Charlie Mason and Jack Benny played in this barn for barn dances. The owners raised hogs here for sixty years.

Wauconda Township. In 1901, the original barn on this property was blown down in a tornado. The barn was rebuilt in 1902, recycling the old beams from the original. The land had been settled in 1845 by Essie and Lloyd Fisher. Vaulting on the new barn utilized modern plank framing techniques.

Wauconda Township. The original barn on this Maitzen family farm burned down around the late 1800s. This one was built in 1900, incorporating styles of the day: a slightly smaller size, with a higher roof pitch, but still utilizing the soil banking and gabled roof of old. Other additions were made to the farmstead in the 1930s.

Wauconda Township. The barn was built in the early 1900s for a small dairy operation. The opening in the upper lofts could be raised and lowered through a pulley system.

5

OTHER FARM STRUCTURES

SOMETIMES CALLED OUTBUILDINGS, MOST FARMS included a variety of structures that were supplemental to the main barn and farmhouse on the farm. These outbuildings were typically single-use structures, but vital to the operation of the farm. As time went on, the number of buildings required fluctuated according to the needs of the generation in which they existed. The single use of many of these buildings restricted their usefulness during generations where they were no longer needed, and proved to be a reason for their demise. Those that were made from concrete or solid materials tended to withstand time simply because they were more difficult to remove.

One of the most significant contributions to farm architecture was the upright wooden silo. Initially, silage, favored for dairy cows, was stored in horizontal wooden bins or cellars built next to the barn or underground. The silage, made up of all of the cornstalk chopped, stomped, and fermented, was preferred for improved milk production from the dairy cows, enabling year-round milking. What was needed was a storage system that would allow loading of silage from the top, fermentation in the middle, and unloading from the base without rodents infiltrating the bin. The first upright wooden silo was built in McHenry County, Illinois, by Fred Hatch. This system of silage handling revolutionized the storage of silage on the farm and freed up a great deal of space previously designated for the production of silage. The trend caught on quickly and was embraced around the country in a relatively short period of time, although some of the farmers resisted this innovation as tales of fermented silage and drunken cows followed the new invention. The improvement of silos with the use of concrete around the turn of the century caused dairy farmers to reconsider their previous hesitation.

By the early 1900s, it was obvious that where there was a silo, there were dairy cows. Initially the designs were entirely wooden with a cone top, but they evolved over the next fifty years, until the final improvement arrived by the late 1940s in the form of blue and white fiberglass

Benton Township. This milk cooling shed at the Strahan farm now functions as a playhouse.

bonded to sheets of metal. Called the Harvestore, this fiberglass silo was first exhibited at the Wisconsin State Fair in 1948. It was airtight to prevent spoilage, freezing, and rust formation. Today the Harvestore, which is very expensive, characterizes a serious and prosperous dairy farmstead.

Dates of the silo can usually be determined by the materials used for the creation. Wooden silos (of which there are few remaining due to rot) are from the late 1800s to 1900. Concrete and cinderblock were experimented with from 1890 through 1920. Tiles and bricks were tested around the 1920s to 1940s, and the large steel Harvestores were widely distributed from 1950 onward. Many farms have a variety of silos, and smaller operations kept the older silos in spite of improvements, simply because the old method was efficient and already in place.

Benton Township. A small milk cooling shed of square design. The stenciling is a new motif.

Also vital to dairy operations were milk houses. The milk house began as a spring cooling shed, where spring-fed pipes would drain into a concrete or stone bin, and milk could be stored and kept cold until taken to market. These spring rooms were not connected directly to the barn, though they were usually close by. These rooms were also referred to as wash houses because milk cans were cleaned here as well. By the early 1900s, milk houses were recommended and encouraged to be built near the end of the milking parlor. The rooms were prohibited from being located in or near the stables, to avoid contamination from other animals. Near the end of World War II, milk houses were officially required and inspected. They were whitewashed and used for brief storage as per the requirements of many state laws. By the 1960s, improvements in milking systems directly pumped the milk from cow to vat, eliminating the need for the storing sheds. After this time, milk houses were converted to work rooms, utility rooms, and tool sheds, or abandoned altogether.

Corncribs, like the silo, began as an underground or close-to-the-barn storage system, evolving over time into those cribs seen today on farms. The original design was improved upon by the pioneers, who created concrete foundations sometimes measuring several inches in depth to discourage rodents. The lateral walls and space between the siding allowed ventilation to dry the corn thoroughly. Most popular was the double corncrib, which provided space in the middle wide enough to allow a wagon to drive in and unload corn into either side of the crib. Heights of the early cribs were designed to accommodate the reach of hands that would be loading the crib. For tenant farmers, the crib was favored because its design meant that it could be easily subdivided to separate crops.

Around 1890, oak slat and heavy steel wire crib kits could be ordered from Montgomery Ward and Sears Roebuck. Experiments in concrete and tiled cribs took place around the 1920s and 1930s, due to the shortage of wood and the need for more inexpensive products. Trends in style tended to follow the roads that traveling salesmen took. Solid concrete was not as favored as the concrete blocks with ventilation slots to permit more drying of the crop, but the drawback in any concrete construction was its propensity for absorption of moisture, encouraging mold growth. The invention of the solo-standing wire-sided corncribs for corncobs, combined with the building of granaries for grain storage, led to the elimination of many early-style corncribs.

The design of the granary mimics the corncrib on a much larger scale. Experiments in a variety of granary building materials took place mostly during the 1900 to 1940 time period. The goal was to find materials that were fireproof, while allowing for ventilation and adequate drying. Most experimental granaries were made from steel, tile,

Ela Township. This unusually designed barn was an ice house located near the rails in Lake Zurich. The building originally had two walls with insulation between them. The various doors would allow limited access to the large blocks of ice. From this location, ice could be stored until the trains came to pick it up for distribution to other locales. The windows were added in later years.

Ela Township. Built in 1941 and equipped with a modern grain elevator, this granary was owned by the Paris family and shared by the area's farms. Note the extra lateral supports on the front of the granary. When full, these supports would prevent the framework from bursting. To the left is a turn of the century design with cone top. This was built next to the original barn, which was destroyed in later years.

and concrete. Often, the granary was shared by entire neighborhoods. The capacity of early wooden granaries can be indicated by the reinforcement of the sides, with wooden slats placed diagonally to support heavy loads. Invention of the ear shelling and corn husking machines virtually wiped out the need for granaries and corncribs after the 1950s.

A note about the grain elevator. Originally the cup elevator was designed and patented by Oliver Evans in 1785. It was not used successfully for commercial grain transport until 1843. Actual acceptance and widespread use of the elevator did not take place until the late 1800s. At this time, the "grain elevator" was utilized on smaller farms as well as in large commercial grain and flour mills. This system used cups, mounted on a continuous chain or belt along a pulley system, that were lifted and tipped to haul products higher into the building. This system permitted a much larger building to be built, since it was not limited to hand-loading height.

Other smaller outbuildings included ice houses, which required drainage from the bottom, protection from sunlight, and double-insulated walls. The older ice house was built partially below the ground and had two walls of stone, with a gap between to fill with hay

Above and facing page: *Ela Township. The first photo is of an old smokehouse and an old outhouse. The stone and masonry construction of the smokehouse indicates an early date. (Probably close to the 1860 printed on the side.) The second photo shows the barn's milk cooling shed wedged into the incline of the soil banked barn.*

or sawdust to increase insulation. The roofs had extended eaves to provide additional protection from sunlight. Later ice houses, built from the late 1800s to early 1900s, were larger and above ground where they could be used by several neighbors. These buildings were identified by a lack of windows, and doors of varying sizes located in odd places around the building. There are few remaining ice houses, and those still in existence have been altered considerably.

Chicken coops were another type of outbuilding designed for specific use. These buildings were distinguished by their long, low, rectangular design. Plenty of light entered through numerous windows, and it was subdivided into "apartments." Nests and roosts were included in the apartments with a run on the outside acting as an exercise area for the chickens. Accessibility for cleaning was considered a must; there was generally a door on each end of the length of the building, with an aisle large enough for a man to walk from one end of the coop to the other. Coops were designed for large chicken populations, rather than just a few for family use that would normally be housed in a small hutch.

The recycling of many of these buildings has not been as easy as the main barn. Because of the specific purpose of these buildings and their unusual configurations and shape, these structures are not as easily altered for newer uses, and thus many have been destroyed. Most remaining outbuildings are there because they were built from concrete or stone and therefore difficult to remove.

Ela Township. A popular silo cover style after the cone shape of the early concrete silos. Constructed of wood, many of these did not survive the passage of time. The latter design of the dome cover in aluminum is the most widely accepted and durable silo top design.

Ela Township. The granary on a Schwermann family farm. The construction was different from the Paris family farm, although they were probably built around the same time.

Fremont Township. Corncrib still in use in Mundelein. There are few farms still functioning in the county. This family farm still feeds a large family in one of the few remaining rural areas. The elongated corncrib was designed with clearance for a tractor to drive through.

Fremont Township. From the same farm, a view of the chicken coop which would have included an outdoor pen. The hens would exit from the small doors.

Fremont Township. Another local granary. The advantage of the large granary was that several farms could store their grain, and it would be separated into sections of the granary. The large size indicated the advanced use of the grain elevator that lifted the grain to the extended height, dropping it in the appropriate bin. This granary was on the Krop estate. To the right is the dome top silo. The dome top was a later design common on silos after the cone design.

Fremont Township. Ribstone silo manufacturer of poured concrete and steel ribs. Popular around the early 1900s. This one was constructed around 1920.

Facing page: *Fremont Township. A corncrib with a drive-through on the Diebold family farm.*

Fremont Township. Granary on the Neilson brothers' farm. The interior view shows the electric grain elevator as well as the size of the scoops that lifted the grain for deposit in the top.

Fremont Township. The old smokehouse at the Krop estate. The smokehouse dates back to the 19th century farm owners.

Libertyville Township. A cone top early silo built in the late 1800s. The cooling shed was spring fed. The extended bays to the left of the barn were common add-ons allowing for expanded herds or storage.

Libertyville Township. Granary modified for use as a stable. Attached to the granary and extending the stable is a pole barn of this era, constructed of corrugated aluminum with a simple wood frame.

Libertyville Township. Full-O-Pep Farm spring house, quite substantial in size. Eggs would be stored here until taken to market.

Newport Township. On a chicken farm, an all-purpose shed. This building at times could be used for slaughter, storage, or as a workshop, and the vent might suggest a separate area for smoking meats.

Grant Township. A small cooling house built around 1900 on the Levi Waite farm. The shed was separate from the barn, but near the milking parlor.

Newport Township. This cooling shed or milk room was attached to the barn at the end of the milking parlor. The white interior was regulation, thought to promote less mold and representative of more sanitary storing conditions.

Vernon Township. Two smaller structures on a farm dating back to the 1850s. These buildings were built in the early 1900s for a storage room, a workshop, and a garage.

Wauconda Township. An example of a cinder block granary with vented blocks. Debate arose over these blocks because they tended to absorb more moisture than originally anticipated. The wooden roof and interior view capture the unique design of the interior and the spiral roof.

Warren Township. Similar granary without the ventilation. There are few of these designs in Illinois. These buildings tended to be along rail lines and/or salesman routes.

Warren Township. A cinder block cooling shed in Grayslake.

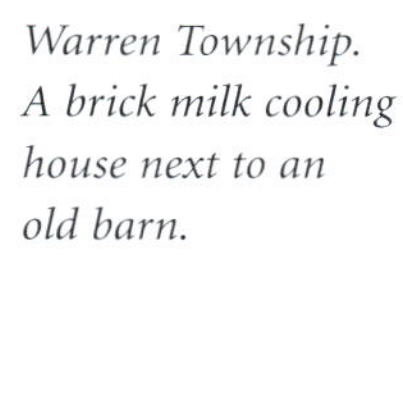

Warren Township. A brick milk cooling house next to an old barn.

6

1930–PRESENT

IT WAS INEVITABLE THAT THE CITY WOULD SPRAWL TO the suburbs. By 1930, more families were moving north from Chicago to enjoy the countryside, and the land was cheap once again due to the declining market. The Great Depression was taking its toll on all aspects of life. The stock market crash of 1929 altered the terrain of the gentlemen farmers. Specifically, men like Samuel Insull, who owned thousands of acres, were forced to sell their land. In Lake Forest, estates were splitting, mansions were demolished, and several homes were built where previously there had been only one. Regular farmers also lost business markets, industrial jobs were harder to find, and labor unions were trying to muscle their way into the suburbs. This resulted in a farming community in transition. Farms were sold. Families, split with the industrial age, were reunited with children now looking for work. Tenant farms increased as ownership of land shifted from ownership by many to ownership by those few who could afford the property.

There were positive changes. Schools, once small one-room buildings, were beginning to consolidate and provide better educational facilities as populations increased. Businesses provided alternative livelihoods to incoming city folk. The Lakes Region continued to attract resort crowds seeking relaxation in the warm months. Summer cottages were converted to year-round homes, and the first hints of modern housing developments were appearing. For farmers, this meant more local business. During this time, standards and regulations challenged the average farmer.

Competition for business in these new neighborhoods was fierce among the dairy farmers who remained. The North Shore and suburban areas increasingly sought fresh milk delivered to the family door. Many farms no longer needed to concern themselves with delivery to large cities when business was right in the neighborhood. Small, organized dairies were formed to offer the best milk in the area.

Lake Villa Township. Vintage 1931. In 1943, the cooling pit was built, although the original land was settled as far back as 1864.

Benton Township. In 1929, a tornado in the township took down these two barns. They were both rebuilt in 1930 and designed very similarly. The property of the first barn was purchased by the Leable family in 1918 for $13,500. The family of carpenters had four boys who fought in World War II, and all four returned home afterwards.

Regulations for these operations were strictly enforced, and it was necessary for farms to consolidate in order to keep up with the times.

By the 1940s, farmers and industry were in conflict. Bob Hill, who operated the Ferry-Hill Dairy in Long Grove, spoke of the competition in his journals. "The Federal Milk Marketing Administration, a government agency, began watching dairy farms closely. They were trying to run the show. And then there was the union flexing its power over our operation. My drivers didn't want to join the union. They threatened to dump our trucks and cause other disruptions. I was told that one of the tricks they used to discourage people from buying from a certain dairy was to spoil the milk by using a hypodermic needle to squirt kerosene through the paper cap. When the big trucks came in from Wisconsin, they'd shoot holes in the tanks with machine guns and cause all the milk to run out."

While the competition between dairies was fierce, there was a growing interest in using the old farms for horse breeding. Farms that were out of business, but still possessing the old barns, were ideal for horses even if they only possessed fifty acres or less. The trend in horse breeding increased. The wealthy hired prominent architects to modify the old dairy farms into horse training centers. By the 1950s and 1960s, Lake County had become one of the most prominent horse breeding areas in the country. Today, some people estimate that there are more horses per capita in Lake County than in any other county in any other state.

The decline of the dairy farm and booming suburbia was in full swing by the 1950s. Cook County was nearly completely developed, and while people still considered Lake County to be semi-rural, there also were a growing number of individuals who foresaw the eventual development of the entire county, if proper steps were not taken to protect it. By the late 1960s, land was being set aside in the form of the Lake County Forest Preserve District, for the sole purpose of preservation of open space.

Ela Township. The closing on the purchase of this farm was December 7, 1940. In the early 1940s, the barn was soil banked. Always enjoyed as a family farm, this barn sits alongside a small pond and apple orchard.

Farming continued to change in the next few decades. More people, improved transportation, and a healthy economy drew families and businesses from the city to the country. The 1970s were a strong time for balance, preservation, and community growth, but not for farming. The number of farms was decreasing with each new generation. Dairy production passed from the Midwest, as demand increased nationwide. Dairy products were now produced in areas where the climate was more temperate and milking could take place year-round without the same needs for the large old barns.

As the national economy fluctuates, urban sprawl does, too. Development of open land today, just as in the early days of Manifest Destiny, is the nature of humankind. Taxation and property values have pushed the small farmer to the limits of his property, and the lack of interest on the part of the younger generations has contributed to the disappearance of the small family farm.

BARNS OF THE MIDDLE TO LATE 1900s

Barns constructed in the 1930s were single-use in nature. Most were smaller family barns, and many were built for horses. These barns stored hay in the lofts and animals on the main floors. They were typically simple in design. One indication of barns built during the 1930s is the roof pitch. To accommodate larger hay storage, the pitch was much higher. Some call this design the Dutch gambrel roof.

There was a large movement to modify the early dairy barns and create horse barns that were attractive as well as practical. Generally on estate farms or belonging to owners of riding stables, these barns were picturesque in their design as well as their aesthetic appearance. The facade was one of the "ideal" farms, without the coarse dirt and roughness of real farms.

By the 1950s the first pole barn was designed. It was built easily by a few men, sold in kits, functional, inexpensive, and practical in its design. Very few barns using old techniques were ever built again after the introduction of the pole barn, except by land managers seeking to recreate the style of the early barns.

Ela Township. Updated from the old carriage-style barn to a horse barn. This building underwent many changes over its history.

Ela Township. Property rumored to belong to Al Capone or his family. Some people say that Capone owned several horse farms in Lake County. Of several already destroyed and developed into other properties, the stories remain just myth.

Fremont Township. Cricket Hollow Farm. This small single-use dairy barn with brick chimney was built around the 1930s. Cricket Hollow is now owned by the Lake County Forest Preserve District.

Fremont Township. One of two contemporary round barns in Lake County. These barns were built in 1998 as horse training centers. The round barn design dates back to the 1820s, when the first round barn was built in Pennsylvania. The allure of the round barn is almost mystical. Some felt that the practical design was excellent for animals. Others felt that the round barn protected the family from evil spirits collecting in the corners. In Lake County, there was a round barn on the Cuneo estate in Libertyville at Hawthorn Melody farms. There is a hexagonal barn in Grayslake that is now a garage, and there was a "barn-in-the-round" on the Tempel Farms property that is hollow in the center.

Fremont Township. Built in 1932 for the Hertel family, who owned several successful family farms in the Grayslake area.

Grant Township. An unusual design constructed in the 1930s or 1940s for hay storage. The design is similar to some tobacco farms in Kentucky.

Lake Villa Township. This barn was built in 1934 by Barry Nevelier. The windows in the lower level, along with the roof pitch, are clear indicators of the barn's age.

Lake Villa Township. Originally on the corner of Cedar Lake and Rollins Road, this barn was built in the 1940s by Andy Anderson, who raised dairy cows. The barn was moved to its present location on an Army tank retrieval truck on July 4, 1982. The dimensions of the barn are 135 feet by 32 feet. The new foundation was built in 1982. When the barn moved, the bats that were in the lofts moved, too.

Shields Township. A barn with many different indications of age. Foundations suggest an older barn, although a new concrete pad was added in the early to middle 1900s. Beams were incorporated from old barns to either construct or add onto an existing structure. It is an example of the elaborate modifications made to dairy barns in the 1930s to 1950s, when many were converted for use as horse barns.

Libertyville Township. This barn was built for horses in 1954 by the Norris family. The family maintained a couple of barns between pastures where the horses could be exercised and fed. The framework of the barn is entirely wood with plank framing methods employed in the gambrel roof.

Vernon Township. This property, which is a horse training center, contains several structures varying in age from the old to the more modern 1950s design.

Vernon Township. Built in 1945, this family dairy barn with a gambrel roof design was converted in 1947 to a kennel for unusual breeds of dogs.

Wauconda Township. The layout of the barn suggests stalls for a pig, a few cows, and perhaps a few goats and chickens. Rather than a dairy business, this appears to have been a small family farm.

Vernon Township. This dairy farm, constructed in the 1930s, was typical of the period. The roof pitch and stalls for horses indicate the barn's age.

Warren Township. A small 1930s family barn in the Wadsworth area. Gone are the doors of the threshing floor, which were replaced by single-use needs for a small dairy.

7

OPTIONS FOR THE FUTURE

"Once the family and the animals leave the barn, the spirit of the barn leaves, too, and a rapid decline of the building follows."

Rick Bott, local barn expert

THE QUESTION REMAINS: WHY SAVE OLD BARNS?

Until communities realize the value of preserving historic barns, the barns are doomed. Unless barns are upgraded in their value architecturally with the support of community engineering standards, they are doomed.

The first step towards preservation is education. Educating the public on the viability and value of barns is easier than educating community leaders. The public already recognizes the aesthetic value of barns. This is evidenced by the popularity of nostalgic books, tours, exhibits, and lectures on barns. The bigger challenge is educating community leaders. It is the community leaders who will ultimately decide the fate of barns, based on viability of use as public facilities, or by setting the construction standards that will allow old barns to be upgraded to modern standards. These standards will need to be flexible to accommodate the individual barn, not to force the barn to conform to regulations already in place.

The second step towards preservation is understanding the manner in which barns were initially designed. Early barns were designed to be preserved. A team of horses could be hooked up to the barn, which could then be dragged easily to another location without compromising the structural integrity. Today, utility lines and modern roadways provide more of a challenge to the relocation of barns, but flexibility of the value will prompt the need for experts capable of doing this. Early barns were designed to be easily dismantled and reconstructed. Again, there are experts who specialize in this work.

Cuba Township. Modified in the late 1950s, this barn features an apartment in the main barn and an efficiency in the silo.

Most challenging to relocate are those barns designed in the early to middle 1900s, because the designs of this period lack flexibility. One option is to lay a new foundation and lift the old framework onto the new base. The wood used and its uniqueness will speak for itself in the aesthetic value of the architecture. Some people feel that these later barns are not as valuable historically, and this may be true today, but in another fifty years they, too, will be relics of the past. Preservation includes foresight as well as an appreciation of history.

Protecting the roof and the foundation are of paramount importance in the preservation of barns. Once gaps in the roof allow elements to enter freely or the foundation begins to shift and allows moisture to enter, it is only a matter of time before the building yields to decay. The elements have always been part of the barn spirit, both in its preservation and in its decay.

Antioch Township. This old dairy barn, on part of a golf club, was transformed into a gracious single-family home.

It is up to our generation to rescue the farm spirit. Barns can be reborn to usefulness in our society. Preserved and adapted to modern uses, old barns have attracted positive attention as much for their historical value as for their beauty and unique qualities. They have a chance to thrive as community centers, homes, educational facilities, cultural centers, businesses, animal shelters, and a multitude of other functions.

At the turn of the last century, there were 2,300 barns in Lake County. By the year 2000, there were fewer than 240 viable barns remaining. While it is the nature of man to modernize, we need to respect and include our past in our present expansion. If we do nothing to encourage the preservation of our historic barns, it is only a matter of time before they are gone forever—reminders of where we came from, if only to be seen in old photographs or in the faded memories of our ancestors.

Deerfield Township. Originally built by the Ott family, some of the original settlers of Deerfield Township, this barn was modified in the 1920s for use as a summer home.

Ela Township. Detail of an old family dairy barn converted into a modern living space. The interior framework adds interest to the decor.

Ela Township. Used as a hands-on museum for children, this barn, owned by the Long Grove Historical Society, is used to teach area schoolchildren about the history of the pioneers and their lifestyles.

Ela Township. Originally located on Cuba Road, this 1840s barn was rebuilt by Al Anderson & Son, builders, of McHenry County as a restaurant at the Millrose Brewing Company in Barrington. The reconstruction incorporated the old "ship's knees" of the original barn into the framework for added architectural interest.

Ela Township. An old dairy building now used by the family for storage and parking. Old barns, when maintained properly, store vast amounts of equipment, automobiles, and family valuables.

Ela Township. Moved several times in its lifetime, this barn was modified by Lee Hickox in the 1940s. When Randolph Street in Chicago was being paved, Hickox drove his Cadillac downtown several times to fill his trunk with old paving bricks. He lined his office in the basement of the barn, the length of his driveway, and walkways around his house with the pavers.

Ela Township. Some families construct new barns in styles reminiscent of the old designs.

Top: *Ela Township. This timber-frame barn was given new life as a reconstructed building. The reconstruction was accomplished using a modern-day crane instead of manpower, including the pine bough at the roof peak during construction. This practice was an old German tradition used to express reverence for the longevity of the pine and to give thanks for the wood.*

Middle: *Ela Township. The barn after completion.*

Left: *Ela Township. The community of Hawthorn Woods uses an old dairy facility for their community center and village hall.*

Fremont Township. This barn/house was initially the foundation for an old barn. During the reconstruction, in the 1920s, it was added as siding around the loft area of the original barn to create a great room. Judging from the interior beams, the foundation probably dated to an early settlement or second generation barn.

Fremont Township. Some people utilize modified barns for business.

Fremont Township. Some people dress their barns with the styles of the day.

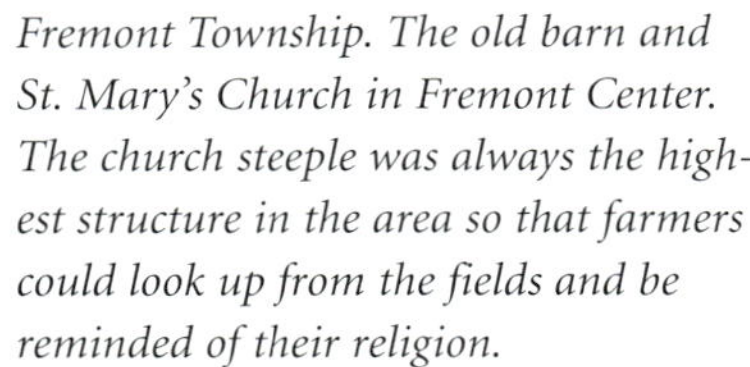

Fremont Township. The old barn and St. Mary's Church in Fremont Center. The church steeple was always the highest structure in the area so that farmers could look up from the fields and be reminded of their religion.

Above and middle: *Lake Villa Township. Originally constructed in Bartlett, this barn was moved and rebuilt as a family home, incorporating all original framework. The chimney was tuckpointed from old foundation fieldstones.*

Lake Villa Township. Some people say that barn lofts were the young men's sanctuary. A church from a barn makes perfect sense. (See also Chapter 3.)

Libertyville Township. This barn was built in 1986 by a father and son working together with old barn parts. It took them fifteen years to collect enough wood from area barns, but they completed the barn just before the son passed away in 1987.

Libertyville Township. The Byron Colby barn at Prairie Crossing. The original one-hundred-year-old barn was on the Byron Colby property down the road. It was dismantled and rebuilt at Prairie Crossing, a conservation community in Grayslake. The barn is currently used as a community center.

Libertyville Township. This was once the last barn in Northbrook dating back to 1873. It was dismantled and reconstructed at its new location.

Libertyville Township. Originally an 1848 timber-frame barn from Wisconsin, it was reconstructed in 1997 by Rick Bott of R&B Enterprises. In Lake County, this kind of adornment on the side of a barn was uncommon. Decoration on barn exteriors was popular among the Pennsylvania Dutch.

Above, right: *Libertyville Township. A home incorporating barn parts for the central framework by Rick Bott of R&B Enterprises. Included are the date and signatures of the original barn builders.*

Right: *Libertyville Township. This structure was built from a Scandinavian barn frame by Rick Bott of R&B Enterprises. The old beams were rehabilitated before they were reconstructed. Scandinavians were known as great shipbuilders. Scandinavian barn ceilings often resembled the inside of a ship's hull, upside down. Today the barn is a family great room.*

Libertyville Township. An old barn, probably built in the 1930s, now serving as a family business.

Newport Township. Some families use their barns to voice their opinions.

Vernon Township. A newer barn with the old church in the background. Modern commercial developments have surrounded these two structures, but they remain as reminders of the past.

Top and middle: *Vernon Township. Before and after shots. A dairy barn originally built in the 1930s, now updated for a modern-day alpaca business.*

Bottom: *Wauconda Township. Here, the Volo Cemetery resides behind an old barn. The tradition of placing the family cemetery on the back one-quarter acre was a settler's right.*

Wauconda Township. Advertising practices on barns trace back to the 1800s. The old Mail Pouch advertisements were originally hand-painted and signed by the artist.

Warren Township. Some farmsteads are used for entertainment like the Tempel Farms complex, where trained Lipizzan Stallions perform for the public every summer.

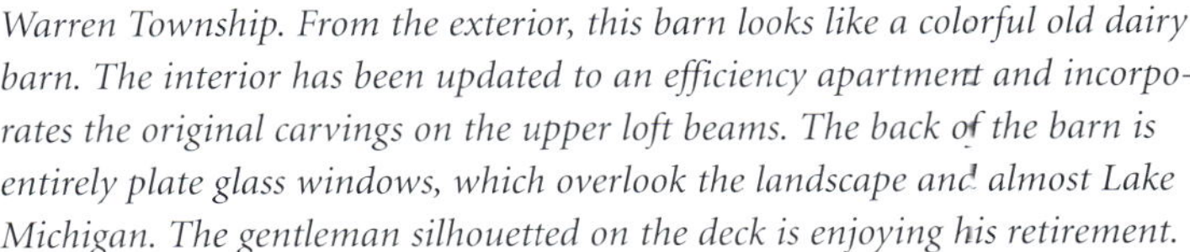

Warren Township. From the exterior, this barn looks like a colorful old dairy barn. The interior has been updated to an efficiency apartment and incorporates the original carvings on the upper loft beams. The back of the barn is entirely plate glass windows, which overlook the landscape and almost Lake Michigan. The gentleman silhouetted on the deck is enjoying his retirement.

Warren Township. The future for our barns? While providing needed experience for fire protection training, every barn lost is a part of our history gone. At the turn of the last century, there were approximately 2,300 barns in Lake County. Today there are fewer than 240 viable barns remaining. This was the Faulkner brothers' farm. The original farmhouse dated back to the first settlers, the Stafford family, who aided those involved in the Underground Railroad and who contributed a school for neighboring children. The original barn was removed to build this modern arch roof barn in 1929 by the Faulkner family. In 1999, the newer barn was destroyed to make way for a new church.

GLOSSARY

Bank barn: A barn built into the side of a natural or soil-stacked hillside. The original German design was for building barns into mountainsides. Originally, there were protected overhangs on the south side. The design allowed for animal storage in the lower level, equipment and crop preparation in the central floor, and hay storage in the lofts.

Bay: The division between any two frameworks in a barn.

Bent: An American tradition, the bent is a section of barn wall that could be built separately and then "raised" later. The early bents measured sixteen feet between posts in post-and-beam construction. In the three-bay design, there were four bents on the long side of the barn. There was generally a 1:2 ratio for length and width of the barn.

Board and Batten Siding: Siding placed with a gap between the vertical planks; after securing, a small strip of wood is added over the gap.

Braces: Support elbows in bents. Earliest were "bent" like "ship's knee" braces.

Continuous Architecture: Begun as a breezeway, but in New England they continued connecting buildings to protect the farmer from the elements. There was a time when this was considered a fire hazard by some communities and outlawed, but eventually it was determined that the farmer could decide to have the hazard or not.

Corncrib: The crib design allows air to circulate and dry the corncobs. They were built late, elevated or with foundation buried, so that rodents would not penetrate the crib. Vats in the side of the crib are open for air to circulate.

Cupola: Designed to improve ventilation and dry hay in the upper lofts, the cupola was first designed around the 1850s. Some farmers preferred this design to lightning rods because they felt that lightning was God's will and the rods were a heathen practice. Early barn fires were blamed upon the lightning, rather than the spontaneous combustion of hay, which was the source of most barn fires.

Door Light: Once a long-hinged board over the door to let light into the barn that was raised and lowered. The second generation of Midwestern barns put panes of glass in this spot.

Double Notch Rafter Seat: A design element before 1875, the rafter seat was a way to attach the rafters to the plates.

Gable: Roof line of simple triangular shape, the standard measure on one-story design was thirty feet.

Gambrel: Named after the hock of a horse's leg. This is a bent rafter line of roof originally designed for the hanging of carcasses at butchering time. Later, it was preferred for maximizing space in the loft. The earliest indicate a European influence in America. It came to the Midwest in the late 1800s.

Girt: Also called tie beam, this is the crossbeam in a bent.

Hand-Hewn Beams: Beams prepared with a broadaxe or an adze. Earliest beams were prepared with an adze, and the hash marks were made horizontally and vertically.

Hay Carrier: Invented by a farmer in Iowa, William Louden, in 1867, the hay carrier ran along or under the ridge beam and delivered and lifted hay front of the ground and into the lofts. Shortly after its introduction, it replaced and redefined the building design of barns.

Hay Fork: Introduced to farmers in the 1860s, the hay fork changed the hay lifting process on the farm. In 1867 it was improved again with the addition of the hay carrier.

Haymow: Haylofts on the sides of an English wheat barn's main floor.

Hex Sign: Colorful symbols on Pennsylvania Dutch barns. Some feel these were designed to protect the barns from evil spirits, but others say they were created "chust for pretty."

Joist: Floor supports that were often round except for the topside, they were the horizontal supports on floors or lofts.

Mortise: The female joint of a timber-frame construction.

Pentroof: Extension on the side of the barn. The original intention was that floor joists would protrude to support a narrow pentroof, which prevented water from draining directly along the side of the building. Some were much larger extensions and later were filled in with walls.

Plate Timber: Top of the bent; this secured the upper ends of the outer posts.

Post-and-Beam Construction: Framework utilizing large beams supporting heavy posts, joined together with wooden pegs in mortise and tenon joints.

Ridge Board: At the roof peak, the ridge board is the support that rafters are connected to. The ridge board was not used in the Midwest until later in the 1870s, particularly after the mass production of the hay carrier, which was attached to the ridge board.

Rough-Hewn Beams: Beams prepared with an axe, but only to prepare the joints and to barely clear the bark off of the beam.

Round barn: First round barns date to the early 1800s. Originally, the religious theme of a circle was the motivation for this design. The Shakers and the Quakers in particular favored this design to keep the devil from the corners. Round barns were an influence on the design of silos.

Salt Box: A roof design of pioneers from England. The long side of the design faced the north, while the higher exposed section faced the warm south. This was to cope with the American weather. Many later barns had the look of the salt box after lean-to additions were added to the barn.

Shakes: The shingles on roofs. Originally, they were made from split logs.

Silage: The most favored feed of dairy cows. The most efficient use of the cornstalk, since the entire stalk is used in silage.

Sill: The framework that rests directly on the foundation to form the substructure for the floor, walls, and roof of the barn.

Silo: The first upright wooden silo was built in 1873 in McHenry County, Illinois. The silo was once just a "hole in the ground." Formerly, silage was ditched or stored in horizontal bins. The first stone silos were created because wood was becoming more scarce. Initially, they were extended as far into the ground as they were above the ground. Where there is a silo, there was, or are, dairy cows.

Springhouse: Built above the ground, the springhouse fed water from underground streams into the concrete pits so that milk would be stored and kept cold.

Tenon: The male joint of a timber-frame construction.

Three-Bay barn: Originally an English design, the three-bay barn was ideal for wheat production and crop storage. The single-story design had a central floor for threshing, with haymows on each side.

Threshing Floor: In the middle of the three-bay English barn, the threshing floor had a door on each side so that after the winnowing, the doors could be opened. The air would blow away the wheat chaff, and the threshold would keep the wheat behind.

Threshold: A small board at the entrance to the doors of the threshing floor that would hold back the wheat from blowing away.

Tie Beam: Also called girt, this is the crossbeam in a bent.

Timber-frame: Timbers (in the early years hand-hewn) that were framed together into bents, and joined together with mortise and tenon joints and wooden pins. Timbers were used to build the frame of a barn.

Tip Window: The window near the roof peak on the side of barns. These windows were extras, brought from an old home or left over after building a new farmhouse. They allowed air and light into the upper lofts of the barn.

ACKNOWLEDGEMENTS

Special thanks to the sponsors, without whose help this book would not have been a reality.

Vein Care Associates
The Liberty Prairie Foundation
MotoPhoto at Woodland Commons
Red Oaks Fine Furnishings
Andee Hausman of Remax Realty
Mom and Dad
Rhoda Martin

Special thanks also goes to the following people who helped provide background information, support, referrals, and other aid:

Bryan Perona of Country Craftsmen
Diana Dretske of the Lake County Museum
Rick Bott of R&B Enterprises
George Von Haegan
Al Westerman
Clayton Brown
A. J. Doherty
Art Miller of Lake Forest College
Dr. Cathy Bird of Midwest Archeological Research Services
Nancy Quintana
Marie Roth
Tom and Tori Trauscht
Mike Roach
Susan Benjamin of Historic Certification Consultants
Lake County Land Conservancy
and
All the barn owners in Lake County who shared their stories, their barns, and sometimes their cookies with my children.

In order to preserve the privacy of the current barn owners, the barns documented in this volume have been listed by their townships. The township form of government was adopted by the early settlers as a way to organize municipalities dating to the earliest colonies in America. We still use the township designation today, although most residents do not realize this.

Disclaimer: Determining the exact age of old barns is a lot like detective work. It was very helpful that some builders dated the rise of the barn, but most of them did not. The earliest barns are sometimes easy to identify because of the rough-hewn beams and lack of milled lumber, but even this is not a guarantee. One old barn in Lake Zurich had old-looking lumber and, thankfully, a note buried in the wall indicating the facts associated with the barn building. As generations grew up and merged their resources and cultures, their barns reflected those merges. Most of the information compiled for this book was based on family stories, old county plats, an old Prairie Farmer Directory, old county records, and books by earlier researchers, who do not always agree on data. One hopes that I am as accurate as possible, but it is probable that new information will be uncovered to retell a few of these tales. This record of history within tells about the farmers who were the foundation for what we see in Lake County today. They are our agrarian roots. These were the families that shaped our history through their hard work, courage, humor, and strength.

BIBLIOGRAPHY

An Architectural Album: Chicago's North Shore. Junior League of Evanston, Inc.: 1988.
Benson, Tedd. *Timberframe.* Taunton Books and Videos: 1999.
Berg, Donald J., AIA. *American Country Building Design.* Sterling Publishing Co.: 1997.
Brown, Clayton. *A Little Bit of History.*
Caravan, Jill. *American Barns.* Todtri Productions Limited: 1995.
Cook Memorial Library
Dretske, Diana. *What's in a Name? The Origin of Place Names in Lake County, Illinois.* Lake County Museum: 1998.
Ebner, Michael H. *Creating Chicago's North Shore.* University of Chicago Press: 1988.
Encyclopedia *Encarta*
Endersby, Eric; Greenwood, Alexander; and Larkin, David. *Barn.* Houghton Mifflin Company: 1992.
Grayslake, A Historical Portrait. Grayslake Historical Society: 1994.
Halsey, John J., LL.D., *A History of Lake County, Illinois.* Roy S. Bates, Publisher: 1912.
Halsted, Dr. Byron D. *Barns, Sheds and Outbuildings.* The Stephen Greene Press: 1977.
Howe, Nicholas. *Barns.* Barnes and Noble Books: 1998.
Klamkin, Charles. *Barns, Their History, Preservation, and Restoration.* Bonanza Books: 1979.
Lake County Discovery Museum
Lawson, Edward. *A History of Warren Township.* Warren-Newport Public Library: 1971; Warren Township Historical Society and Warren-Newport Public Library: 1997.
Mitchell, James. *The Craft of Modular Post & Beam.* Hartley & Marks Publishers: 1984, 1997.
Mullery, Virginia. *Lake County, Illinois.* Windsor Publications, Inc.: 1989.
Noble, Allen G. and Wilhelm, Hubert G. H., eds. *Barns of the Midwest.* Ohio University Press, Athens, Ohio: 1995.
Park, Virginia. *Long Grove Lore and Legend.* 1978
Pictorial History of Ela Township. Ela Historical Society.
"Prairie Farmers Reliable Directory of Farmers and Breeders of Lake County." *Prairie Farmer.* Chicago: 1917.
Scott, Donald H. *Barns of Indiana.* Donning Company Publishers: 1997.
Sloane, Eric. *An Age of Barns.* Funk and Wagnall's Publishing Company, Inc.: 1967.
The Past and Present of Lake County, Illinois. Wm. Le Baron & Co., Chicago: 1877.
This Old Barn. Reiman Publications: 1996.
"This Old Barn"
Tregillis, Helen Cox. *River Roads to Freedom.* Heritage Books, Inc.: 1988.

SURVEYS
1861, 1885, 1907, 1936, 1954, 1976

INDEX

ABOUT THE AUTHOR

Nancy Schumm-Burgess

Writer/photographer Nancy Schumm-Burgess is the author of *Gardens and Other Sanctuaries in Long Grove, Illinois*, and *Hearts Full of Compassion.* Burgess has been researching and photographing the remaining historic barns in Lake County since 1997. She is a former board member of the McHenry County Historical Barn Preservation Association, former chairman of the Illinois State Historical Society Save-Our-Barns Committee, and founder of the Lake County Save-A-Barn network. A history buff, Schumm-Burgess has been writing historical articles in Lake County since 1994 for numerous publications. As a photographer, her photographs have been exhibited in several museums throughout Illinois, in local publications, and on the Oprah Winfrey Show.

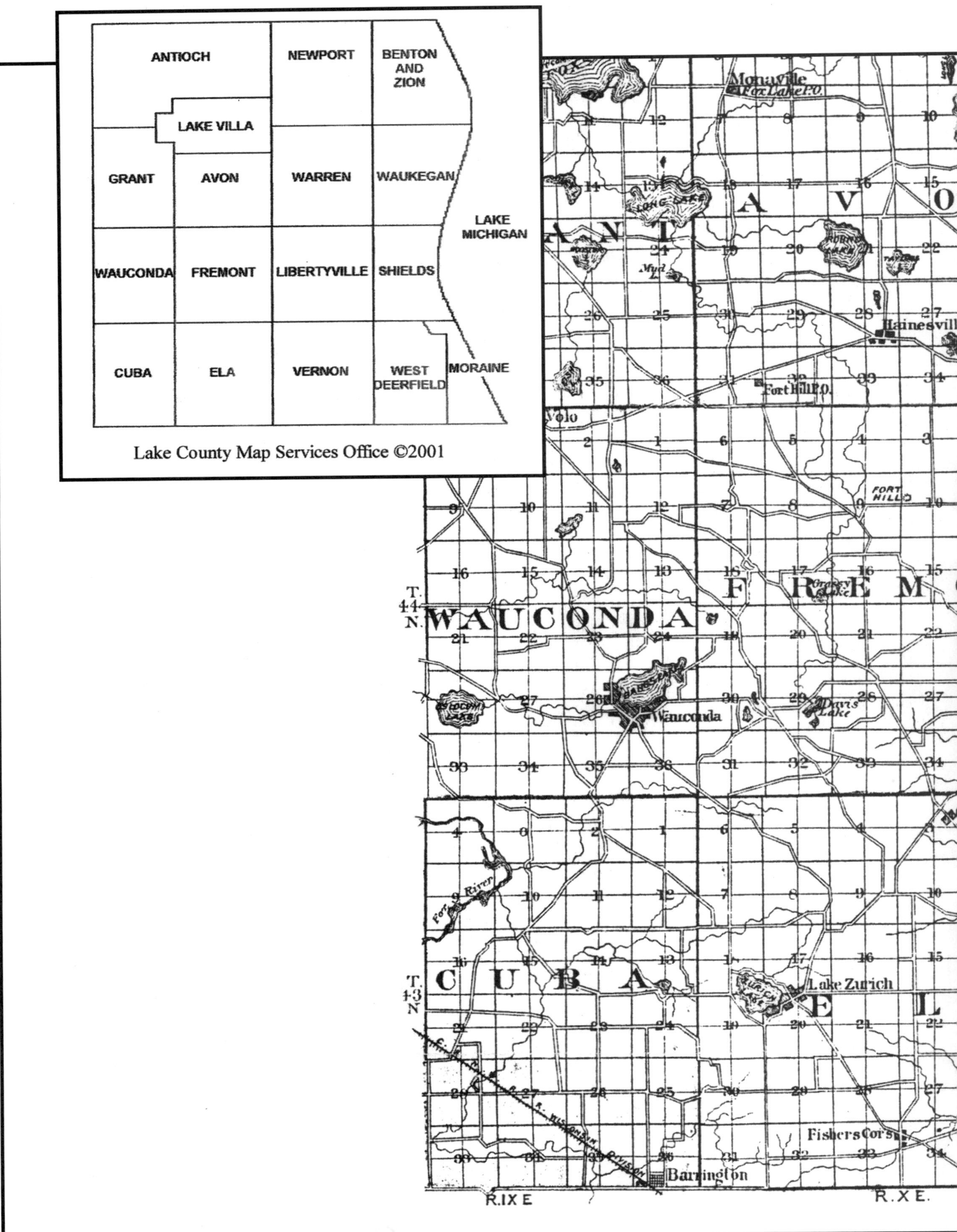
ANTIOCH
NEWPORT
BENTON AND ZION
LAKE VILLA
GRANT
AVON
WARREN
WAUKEGAN
LAKE MICHIGAN
WAUCONDA
FREMONT
LIBERTYVILLE
SHIELDS
CUBA
ELA
VERNON
WEST DEERFIELD
MORAINE
Lake County Map Services Office ©2001
Monaville
Fox Lake P.O.
Long Lake
Mud L.
Hainesville
Fort Hill P.O.
Volo
Fort Hill
W A U C O N D A
F R E M
Wauconda
Davis Lake
T. 44 N.
T. 43 N.
C U B A
E L
Fox River
Lake Zurich
Fishers Cors
Barrington
R.IX E
R.X E.